City of Equals

City of Equals

Jonathan Wolff
and
Avner de-Shalit

Great Clarendon Street, Oxford, OX2 6DP,
United Kingdom

Oxford University Press is a department of the University of Oxford.
It furthers the University's objective of excellence in research, scholarship,
and education by publishing worldwide. Oxford is a registered trade mark of
Oxford University Press in the UK and in certain other countries

Published in the United States of America by Oxford University Press
198 Madison Avenue, New York, NY 10016, United States of America

British Library Cataloguing in Publication Data

Data available

Library of Congress Control Number: 2023942633

ISBN 9780198894735

DOI: 10.1093/oso/9780198894735.001.0001

Printed and bound by
CPI Group (UK) Ltd, Croydon, CR0 4YY

To Elaine and Yifat

Preface

Consider the photo on this book's cover. It was taken in 2004 by the Brazilian photographer Tuca Vieira and immediately caused a sensation. Vieira's goal was to show Brazil's 'brutal inequality', as he explained in an interview in *The Guardian* (Vieira 2017). It shows part of the Paraisópolis favela in São Paulo, side by side with Morumbi, a very wealthy neighbourhood, in the southwest of the city. Obviously, upon seeing this photo the first thought that comes to mind is the shocking material gap. On one side people enjoy luxury, economic stability, and the prospect of high-quality leisure time (even if they are not seen there to be using it). On the other side, people are struggling just to get by, desperate to find jobs, and in the heavy Brazilian rainstorms their crowded and hastily constructed homes are liable to flooding.

Nevertheless, there is another aspect that appears on a closer look. While the tennis court and the swimming pool in Morumbi seem empty, and no one is using those curious water facilities on the balconies, in the favela people can be seen walking, and stopping to talk to each other. There is a sense of community in the favela, but in the high rises, Vieira, it seems, wishes to show both a colossal waste of resources and the alienation of people from each other. But what also becomes evident, and strikes hard, is the line that separates the two neighbourhoods in a single city. It is a wall, that distinguishes, excludes, and defines people living more or less next door to each other as us and them. This seems to us nothing but humiliating. When we showed this photograph to our students the most common reaction is that it is deeply depressing that people can become habituated to such indifference to each other. Although this photo depicts just one small part of one vast city, what it represents is present—overtly or disguised—in cities all over the globe.

In the interview Vieira talks about how he feels about what the photograph depicts, taking into account many issues which go beyond material inequality:

> The unjust and brutal difference between rich and poor, inherited from slavery, is in the origin of many other problems—violence, below-par schooling, prejudice and many other issues. Inequality means that someone who is poorer is obliged to work more, so they have no time to study, which impacts on their education. As a result, they are not able to develop the critical and political awareness necessary to transform their own situation. Add to this the absence of any sense of the

> collective responsibility or solidarity among the privileged in Brazil, and you have a closed circle that is very difficult to break.
>
> (Vieira 2017)

In previous work we described this as 'clustering of disadvantages' (Wolff and de-Shalit 2007). Here we want to stress that while the material inequality of the picture is dramatic, Vieira is especially concerned with forms of what are often thought of as social or relational inequality: the physical and symbolic walls that separate and exclude, the prejudices and suspicion which alienate, all of which create a situation whereby members of the same spatial community do not regard each other as equals. But, more encouragingly, there is evidence that within communities such as the Brazilian favelas, despite mental health challenges and the constant threat of violence, strong, supportive and protective, social networks of families and friends develop, even if, sadly, those networks terminate at the favela boundaries (People's Palace Projects and Redes da Maré 2020).

In this book we argue that a city of equals is not defined merely by reference to gaps in wealth or income, but rather by being able to secure a sense of place to all its residents despite economic differences; a sense of place as a member of a city as a whole, and not simply within a very local area such as a very restricted neighbourhood. Focusing on a secure sense of place might appear counter-intuitive. Income equality, it will be said to us, is surely the most important component of equality in the city. This derives from the basic egalitarian intuition that because income and wealth are highly important for achieving whatever goals one has in life, equality in income or wealth are of central concern. We do not deny the importance of material and economic factors. However, as we argue in this book, cities differ from states in many ways, which raises the question of whether equality in the city should be analysed as a type of micro case of equality in the state, as if cities are micro-states. We argue against this view. For example, if we were to rank major US cities in terms of embodying what we will sometimes call 'the egalitarian spirit', in the sense of exemplifying features that egalitarians value, Berkeley California would probably come high on anybody's list. But Berkeley also has very high income inequality, as it attracts both multi-millionaires and (due to its liberal ethos and reasonable weather) those who have nowhere to live but the street. Therefore when the Gini coefficient of 300 cities in the United States was calculated, Berkeley was found to be the tenth most unequal with a high Gini of 0.5438. The lowest quintile in Berkeley receives only 1.42 per cent of the city's total household income, compared to 55.42 per cent received by the top quintile (Knobel 2014). In this sense, perhaps paradoxically, financial inequality in a city can be a sign of inclusiveness, provided, of course, it is not accompanied by spatial exclusion, gated communities, ethnic discrimination, and the like. Contrast Berkeley with a city that has few poor people and very low ethnic diversity. It may do much better in terms of income inequality yet sounds, initially, rather unappealing to those with the egalitarian spirit.

The idea that a city of equals is a city where all residents enjoy a secure sense of place is itself difficult to describe in detail, though our aim is this book is to achieve an explanation by analysing it as consisting of four 'core values' of a city of equals (non-market access to goods and services; sense of meaning; diversity and social mixing; and non-deferential inclusion) and several further themes related to each of these core values. We are writing to articulate the implicit understanding of the values of the city held by those who wish to live in circumstances in which they can regard themselves and others as equals, and that this perception of equality is not simply a matter of subjective feeling but is underpinned and reinforced by the underlying organization and arrangement of the city. We hope that those who share the egalitarian spirit will recognize what we say as a statement of their own values. People who share such values with us, the authors, don't wish to look up at those above them or down at others. They oppose snobbery and servility, and seek ways to include everyone in the material and non-material benefits of the city, rather than blame poor or marginalized people for their condition, or to exclude them from society. In other words, we argue that this egalitarian idea should appeal to those who wish to live in a city that is inclusive and welcoming, regardless of race, gender, age, sexuality, religion, disability, social class, or any other characteristic that sometimes drives people apart. To clarify, in this book we do not set out to argue that people should have the egalitarian sentiment, or spirit (although, of course, this is something we believe and have explored in other work). Rather we want to understand what makes some cities more congenial than others to people with that spirit. This, in turn, may help guide city policy in the future for those cities that aim for equality.

It is often the case that political philosophers and theorists apply a critical approach to whatever they study, and although this book follows that general approach, it is, perhaps, less critical than others may have been. For it is written by two urbanites who are in love with their cities. Jo has spent almost all his adult life living in London, and even as a child never lived more than just a short trip away, and Avner grew up, and has made his own home, in Jerusalem. We might not always like every aspect of our cities, but we love them. Moreover, we both believe that when things go well cities can indeed be inclusive and offer a secure sense of place to all their residents. To borrow a term from the late sociologist Erik Olin Wright (2010) we believe that such a city is a 'real utopia'.

Thanks and Acknowledgements

We would like to thank first and foremost our research assistants, Jakob Tonda Dirksen, Katarina Pitasse Fragoso, Tal Eldar, and Tal Alster, who were extremely helpful not only by conducting many of the interviews, but also in offering critical input and valid ideas. We have discussed the ideas presented here when they were still half-baked, or not even that, in various seminars, lectures, and mini courses, some online and some in cites, including Abu Dhabi (NYU), Athens Georgia, Bern, Durham, Jerusalem (the Hebrew University), London, Lund, Macerata Italy (The Giacomo Leopardi School of Higher Studies), Melbourne (Monash University), Oxford (Blavatnik School of Government), Portland (The Philosophyofthecity association annual meeting), Princeton, Rio de Janeiro, Rotterdam, Sapir College, São Paulo, and Shanghai, as well as during a special summer course we taught in Dubrovnik on behalf of the South East Europe Center for Advanced Studies. We warmly thank the organizers and participants, for their fruitful, helpful, and indeed challenging questions and suggestions. We also thank our students at Oxford University and the Hebrew University of Jerusalem for their comments and questions while we were teaching our seminars in political theory and urban political theory.

Many individuals read parts of the manuscript or discussed our ideas with us. For very fruitful comments, suggestions, and points of criticism we would like to thank Rana AlMutawa, Tal Alster, Daniel Attas, Helmut Aust, Alice Baderin, Jana Bacevic, Nir Barak, Ana Barone, Ori Bechor, Daniel Bell, Galit Binet, Petar Bojanic, Nico Brando, Noam Brenner, Arina Cocoru, Elaine Collins, Maria Dimova Cookson, Jamie Draper, Tal Eldar, Marian Giannotti, Lior Glick, Daniel Guillery, Paul Heritage, James Hickson, Merav Kaddar, Aaron Kaugman, Hazem Kawasmi, Volker Kirchberg, Elizabeth, Kahn, Margaret (Peggy) Kohn, Pilar Lopez Cantero, Hélène Landemore, who actually suggested the book's title, Patti Tamara Lenard, Pedro Lippmann, Pedro Logiodice, Gil Maymon, Marco Meyer, David Miller, Dan Miodownik, Giuliana Pardelli, Katarina Pitasse Fragoso, Rahul Sagar, Yael Shmaryahu Yeshurun, Ron Sundstrom, Bianca Tavolori, Bart van Leeuwen, Paz Yaacov, Nechumi Yaffe, Daniel Weinstock, and Steve Welch.

We also want to thank the three referees whose comments and suggestions were sharp, insightful, and constructive. Authors often complain about the referees, but we genuinely believe that their suggestions helped us to significantly revise and improve the manuscript. Working with Dominic Byatt, OUP's editor, is always a pleasure, and we thank him for making the submission process so smooth and painless. The production team was incredibly helpful. We want to thank Marten Sealby, Phoebe Aldridge-Turner, Raja Dharmaraj, and Susan Frampton.

We also thank from the bottom of our hearts the 182 interviewees for their time, for sharing their world views with us, and for insightful discussions of the questions we posed to them. This book would simply not exist without their generous cooperation.

Part of this research was conducted with the financial help of the Alfred Landecker Programme at the Blavatnik School of Government, University of Oxford, funded by the Alfred Landecker Foundation, and the Max Kampelman Chair for Democracy and Human Rights at the Hebrew University, which we gratefully acknowledge.

Finally, we would like to thank Elaine, Yifat, Max, Daniel, Hillel and Shiri, for their patience, support, and encouragement. We hope we have not been too unbearable while working on our research.

This research, to the best of our knowledge, is first of its kind. Our methodology combines philosophical reasoning with interviews with city-zens in ten cities in six countries. One of the key elements that strongly emerged from the interviews is that in a city of equals a person's ability to enjoy a meaningful life should not be fully determined by their economic resources. We therefore arranged with the publisher Oxford University Press (with the generous support of the funders of our research The Alfred Landecker Foundation and the Max Kampelman Chair) that this book should be published Open Access so that it should be accessible and free to whoever wants to download and read it. We shall be more than happy to read your comments, and given how the world is these days, our email addresses are not difficult to find.

Jonathan Wolff and Avner de-Shalit

Contents

1
Introduction, Motivation, and Methods

1.1. The Research Question

This book is motivated by a simple question. Just as Jane Jacobs asked what makes some parts of cities feel safe and others unsafe (Jacobs 1961), we ask: what makes some cities attractive to people who think of themselves as progressive, liberal, egalitarians (such as us, the authors), and what makes some cities less attractive to such people? Of course, people can be attracted to cities for all sorts of reasons: the theatre, the music, the proximity to the beach or the mountains, the quality of the coffee, and so on. But still, there are some cities that do more to embody what we can call 'the egalitarian spirit', which can be part of the package of features that attracts egalitarians to live there, and it is this we hope to capture. In the words of the title of this book, what makes a city (more of) a city of equals?

It might appear as if we run together two questions: what makes a city egalitarian and what makes it attractive to (self-described) egalitarians? Of course, these are not the same question even if the answers may be connected. So let us clarify: the former is the book's framing and its core; the second is the puzzle that motivated us to do this research. It seems that, almost paradoxically, and some might even allege hypocritically, egalitarians prefer cities where prima facie, there is quite a lot of material inequality. But the main research question is, indeed, what is a city of equals? Another way to put it is to ask what kind of city would be attractive to people who are disadvantaged, assuming that disadvantaged people will find it more appealing to live in a city of equals than elsewhere. This, of course, calls for a theory of disadvantage, which we have discussed elsewhere (Wolff and de Shalit 2007).

Our answer will, in fact, have much in common with Jacobs's own approach, and much less in common with what, to many, may seem the obvious answer: that what matters is the distribution of income and wealth in the city. In fact, it is the lack of correlation between income inequality and attractiveness to egalitarians that fascinates us and drove us to reflect further. According to compilations of statistics, some of the most inegalitarian cities in the world, in terms of income and wealth, and especially in the United States,

City of Equals. Jonathan Wolff and Avner de-Shalit, Oxford University Press. © Jonathan Wolff and Avner de-Shalit (2023).
DOI: 10.1093/oso/9780198894735.003.0001

are the ones egalitarians—rich or poor—would most like to live in. The most obvious example is Berkeley, California, as we mentioned in the Preface, the city of choice for many egalitarians but at the same time a city of staggering financial inequality (Lu and Tanzi 2019; Euromonitor 2013; 2017; Adamczyk 2019). (And as we shall show, this might be because the people of Berkeley tend to be egalitarian and therefore people who are poor or even homeless are attracted to this city.) Other obvious examples are Frankfurt, Germany, and Jerusalem, Israel, which are often thought by locals to be rather egalitarian in their nature, while in terms of income gaps are among the most unequal cities in their countries.

A prior question, no doubt, though, is why focus attention on cities and the policy of equality in the city rather than develop a general theory of equality and apply it to cities? The answer to this question partly emerges from our approach to political philosophy, which we characterize as falling into the tradition of 'Bottom Up Moral and Political Reasoning'. To explain, one dominant approach in political philosophy has been to try to develop theories of equality or justice in abstraction, through methodologies such as a hypothetical contract, or hypothetical auction, or from refinement of a theory in the light of ingenious counterexamples. Once such a theory is developed it can be applied to the state and, perhaps, to the city as well. In contrast, while we recognize that these abstract contributions can enrich our philosophical understanding of justice and equality, when we turn to recommendations for policy we subscribe to the view that political philosophy should begin with understanding the challenges to policy makers, and then bring relevant moral considerations to this understanding (Wolff, 2019b, 9–10). In this particular case we do not merely wish to see how abstract theory can be applied to cases, or how abstract theory can inform public policy, but actually to begin from policies—good and bad—that we observe around the world, as well as the problems identified and, sometimes, the solutions offered, by residents of cities whom we interviewed. From these materials we are inspired to build up a more general theory that is sensitive to real-world achievements. Certainly, abstract theories can also guide us, but our first move is not to jump from the policy challenges to the application of an existing theory, perhaps designed initially for other purposes, but to theorize on the basis of these challenges, and thereby generate an approach tailor-made for the subject matter of a city of equals.

But still, why focus attention on cities? Why not pay attention to equality within a state, which after all has many instruments available to tackle inequality if it so wishes, or perhaps to global justice where problems of inequality are so much more pressing? We agree that these are vital areas for examination and research. They may well be more important, although we do not need to take a stand on that issue here. All we need point out is that state-level and global inequality do not exhaust the terrain.

Generally speaking, thinking about justice, equality, egalitarianism, and even democracy in the city, can take two paths. One is to assume that the city is like a mini-state. It is yet another institution in the state, which behaves according to the same rationale, and therefore should be subject to the same ethical principles. Even the expectations of citizens from the state and its apparatus are more or less the same as the expectations that city-zens (citizens in cities) have from the city and its apparatus. Following such assumptions, when political philosophers come to think about equality in the city they should prima facie rely on theories and principles of justice and equality developed in the context of the state, and perhaps modify them here and there to suit the particularities of the context of the city.

But another path, the one we suggest to take, is to acknowledge that treating the city as a mini state is what was once in philosophy called a 'category mistake'. The city is indeed a political institution, but of a different kind than the state. It is not that the city is like a state, only smaller. In fact, empirically this is not so true. Many cities have budgets which are larger than budgets of many states. This is so even within Europe and within the United States. At the time of writing this book, London's budget equals more or less the budget of Croatia or of Slovenia, and is twice the size of Latvia's, and three times that of Bosnia and Herzegovina. New York City's (NYC) budget is bigger than the budget of forty-three US states, and Chicago's budget is bigger than that of thirteen US states.

Moreover, the city is a different kind of institution, as its rationale differs. First, to put it starkly, the rationale of the state is separations and borders—it is about putting a boundary around individuals and protecting their rights, and distinguishing the state and its citizens from other states and other states' citizens, whereas the rationale of the city is connecting. The city connects city dwellers with each other to create a strong local economy and vivid local community, but just as importantly connects the city with its wider regional environment, bringing in and sending out people, goods, and services in a system of mutual dependence. Second, relationships between city-zens and their municipalities or mayors differ significantly from relationships between citizens and their prime ministers and ministers. Within the city political relationships are often much more intimate than those between citizens and their governments, prime ministers, or ministers. Consequently, also the distribution of power between cities' authorities and their residents differs from that between states and citizens. Mayors and city councils often run their institutions in a more technocratic manner than governments do, being less committed to this or that ideology. The legendary mayor of Jerusalem, Teddy Kollek famously said: 'Gentlemen, spare me your sermons and I will fix your sewers.' Even if Benjamin Barber exaggerates when he writes that 'that's what mayors do, they fix sewers' (2013, 91), it is quite common to believe that mayors are less committed to ideological positions than legislators and governments. Most cities are also limited in their ability to raise taxes and therefore in the way they are able to handle questions of equality. As a result of all this, we argue, city-zens's expectations from cities differ from their expectations from states. The two are rather different kinds of political institutions, and therefore the idea that prima facie we should simply apply principles and theories of equality from the state to the context of the city would be very difficult to do, even if it were the preferred approach (which for us, as should be clear, it is not). It is at this point that we depart from many of the works on equality in the city.

Thus, equality at the level of the city should be discussed and studied separately from equality at the state level. And equality at the level of the city is, we believe, an under-researched area, and one reason for this current research is simply to attempt to fill this gap. Our previous work puts us, we hope, in a good position to make our own contribution (Bell and de-Shalit 2011; de-Shalit 2018; Wolff and de-Shalit 2007; Wolff 2019a). Much has been written about equality and inequality in the state, and, commencing in the 1980s, also much has been written about global inequality or cosmopolitan justice. At the same time, very little, if anything, has been written about inequality in cities.

While Ancient Greek philosophers, most notably Aristotle, did directly discuss the city, and at times some cities themselves have in effect constituted states, as in the case indeed of ancient Greek cities, or Rousseau's Geneva, modern and contemporary political philosophers have, to this point, paid little attention to cities as objects of analysis in their own right. In aiming to address this gap we are not alone, and in the last few decades there has been growing philosophical discussion of the city, including work on the just city, which we explore in Chapter 2. This discussion has been taking place both in philosophy and in allied disciplines. We build upon some of the most important contributions to this literature, as we will show in Chapter 2. But this is still an emerging field with much work to be done. We are pleased that cities are now being put under the philosophical spotlight, though feel that the task for philosophers has only just begun at any scale.

But the full answer to why we have undertaken this project goes well beyond the idea of simply filling a gap in the literature. These reasons relate to what is distinctive about cities, and what makes conceiving and measuring inequality in them so different from conceiving and measuring inequality in the state, for cities differ from states in many ways that make traditional approaches to equality problematic, as we shall argue shortly. But we want to argue that cities are important enough politically and economically speaking for us to devote a whole book to this topic.

First of all, cities are home to more than half the world's population, and are assuming ever increasing political and cultural importance, driving economic prosperity, and dominating global trade (Bradley and Katz 2013a; 2013b). Most of the world's population—around 4 billion people at the time of writing this book—live in cities, particularly in developed and middle-income countries, and in some European countries, a sizeable majority (68–85 per cent) lives in cities. Cities are responsible for nearly 80 per cent of states' GDP. Therefore, Brookings scholars Jennifer Bradley and Bruce Katz (2013a) write: 'The national economy is a network of metropolitan economies.' Cities dominate trade, generate their own wealth, and consequently, find themselves with considerable political power (Barber 2013). As already noted, some cities today spend more than many states. Thus, Parag Khanna (2012) claims, albeit controversially, the twenty-first century will not be dominated by China or India but by cities (see also Barber 2013; David Harvey 2019; and Bell and de-Shalit 2011). Whether or not Khanna is right, it would be foolish to deny the political, social, cultural, and economic importance of cities (Clarke and Gaile 1998).

At the same time, inequality in cities is becoming a daunting issue (Musterd and Ostendorf 2012; Sassen 1999). Economic inequality in cities is growing (Glaeser et al. 2009), perhaps reflecting changes in employment opportunities (Nijman and Wei 2020) and causing political instability (Musterd et al. 2017). The UN-Habitat report of 2016 declares that 75 per cent of the world's cities have higher levels of income inequalities than two decades ago. As King (2011) argues, the sheer scale of urban growth is likely to exacerbate poverty and inequality even more. There are new, brutal forms of deprivation in many cities around the world. Gentrification, rising rents, and the withdrawal of public services push poorer residents to the periphery, to temporary accommodation, or even the streets. According to 'the world city hypothesis' (Friedmann 1986) metropolises are especially prone to extremes of inequality, with dense clusters of poverty adjacent to 'concentrations of the extraordinarily wealthy' (Fainstein 2001) while the middle classes flee to the suburbs (Sassen 1991). (As we will discuss in Chapter 2, these particular class-based housing patterns have changed in more recent years, with mass gentrification in the inner cities being a new factor, but cities remain patterned by class.) Housing is perhaps the most visible sign of deprivation and inequality, from the slums of the developing world, to the shocking clusters of tent-dwellers appearing overnight in some of the world's wealthiest countries. Equal city-zenship and some sort of fairness in opportunity and enjoyment of life, seem absent, even discouraged. At the same time many people report that they are attracted to the 'urban way of life', and any approach that attempts to diminish the dominance of cities feels utopian and out of touch with both the practicalities of contemporary life and the priorities of many people today.

A city's wealth and its internal inequality often correlate (Tonkiss 2017). London, for example, is the United Kingdom's richest city but, in terms of income, wealth, and housing security, its most unequal (Aldridge et al. 2015; King 2011). But how inequality in cities should be measured has rarely been discussed in detail. Although, as we have said, it has been taken for granted that the right measurement is gaps in income and wealth, in many academic works racial or ethnic segregation, especially when it takes a spatial form, is added to the analysis. In this book we raise the question how inequality in cities should be understood, or, put differently, what it takes to characterize a city as what we call a city of equals. We distinguish between the standard accounts of equality in the city, which take distribution of income and wealth as the defining issue, and a city of equals, which, we argue below, pays much more attention to how people feel, what they can do, and how they are treated and regarded by others, thereby embodying what we have referred to as 'the egalitarian spirit'. To make our case we have to explore what egalitarianism (and equality) in cities means and whether it is significantly different from egalitarianism (and equality) in other institutions, not only the state but also the family, or workplace, or other parts of life, although we will not explore these non-state contrasts in detail.

It is also worth pre-figuring a distinction that needs to be kept in mind throughout this book. Although the literature is full of important suggestions about how to *measure* inequality in the city, we find it weaker on the question of *definition* and, therefore, on a firm grounding for the accuracy of the measures. To explain by way of a related analogy, it is common in the poverty literature to measure relative poverty in terms of falling below 60 per cent (or some other percentage) of median income. But this is only a proxy measure. Relative poverty is traditionally defined in terms of whether one is able to be included in society on the same terms as others. A typical definition could be summarized as: 'not being able to do what is normally expected or encouraged in your society' (Townsend 1979), and it is thought that if you have less than 60 per cent of median income you are very likely to miss out on some activities or forms of consumption taken for granted by others. It is the 'missing out', not the income inequality, that captures the essence of relative poverty. Similarly, we can measure inequality by income gaps or commuting time or access to leisure facilities or segregation in the workplace or equality in education, but the underlying question is what is it that we are measuring by these proxies? Here, as we shall set out in detail over the course of this investigation, our core insight is that a city embodies the egalitarian spirit, and is thereby a city of equals, to the degree that it gives each person what we will call 'a secure sense of place', or a sense of belonging to the city, on the same terms as others. How further to specify, and ultimately to measure, this idea of the egalitarian spirit is the central focus of this book.

It is also worth saying that our question is an internal one, not an external one. For the purposes of this project we are interested in equality within a city, and not, for example, relations of justice between the urban centre and rural periphery, or inequality between different regions in a country. These, once more, are vitally important questions (Chauvin 2021). But they do not take up the whole space, or render our study irrelevant.

When reading the account we present over the course of this book, some may think we have captured the idea of a 'good city', or of the benefits of urbanism in general, rather than 'a city of equals'. But nevertheless our motivation is to explore the idea of an egalitarian city. We are, of course, influenced by the benefits of urbanism and exploring the idea of a city of equals inevitably raises the questions of the benefits of urbanism and how they are distributed among all city dwellers. A truly good city, in our view, is one that provides the benefits of urbanism for everyone who lives within it. It is therefore, in that respect, a city of equals.

1.2. What Is a City?

First and foremost, we need to explain what we mean by a 'city'. The city is an elusive concept (Parnell 2015). In the contemporary world providing an account of the city is not a straightforward matter. David Harvey summarizes Max Weber's defining characteristics for early 'occidental cities' as: 'A fortification; a market; a court of its own and partially autonomous law; a distinct form of association and partial autonomy and autocephaly' (Harvey 2009 [1973], 305). Times move on, and factors around partial independence, size and organization, and identity have become the critical factors. Geographers often focus on spatial issues referring to a continuous district of settlement,[1] urban sociologists focus on individuals or groups of people and on population density, lawyers and political scientists often refer to a single jurisdiction,[2] and urban economists often refer to the scale of economic activities. Some define the city on an institutional basis (referring mostly to its jurisdictional borders)[3] and some on a functional basis.[4] So let us at the outset offer our definition of the city.

We take the city to have four primary components. Cities are, first of all, *institutional*, meaning that they have a single jurisdiction. This is a typical factor, not a necessary condition. London has thirty-two boroughs (plus the City of London, which is a ceremonial county and local government authority), each of which functions as a jurisdiction, in addition to a central authority. Yet between 1986 and 2000 London did not have a mayor or a city-wide leader. However, this is clearly an anomaly, and it is natural to regard a city as highly typically having a political infrastructure. It would also be possible to regard each of the London boroughs as sub-city. Many are of the size and complexity as entities regarded as a city. At this stage we are neutral on whether they should be so regarded.

A second aspect of the city is that it is *densely populated*, at least relative to its surrounding area. Of course, there can be parts of cities that are less densely populated. There are parks, some cities such as Berlin and Paris even have farms within the city boundaries, and in some cases the enclaves of the highly wealthy are much less densely populated. But the jurisdiction as a whole will typically be more densely packed than the surrounding area, though we are also aware that there are conurbations in which cities merge into each other with no remaining less-populated space in between.

Third, cities have their own *cultural-political identity*. They encourage civicism, or pride in the city (Bell and de-Shalit 2011), and facilitate and encourage an intense 'urban' way of life involving particular forms of commerce, transport, leisure activities, and so on. Typically, each city has its own

[1] Although in some cases, such as Paris, the metropolitan area includes some rural areas and some areas with lower-density settlements that are not really 'urban'. Rio de Janeiro even has rain forest and mountains in the city centre. We are primarily concerned with what city dwellers regard as the urban core of their cities, because we are interested not only in the city as a form of government, but more so in the impact of the urban way of life on inequality.

[2] UN-Habitat 2009. Some theorists claim that cities are often governed by a system of formal and informal relationships and that the formal government of a city could never really achieve its goals without collaboration with civic elements and interests, and therefore a city is often governed by informal relationships. For more about the 'urban regime theory' see Stone (2006), and for a critique see Smith (2013). About how we reflect upon cities in a global age see Derudder et al. (2011).

[3] See, for example, Briffault's (1996) definition of the locality as a territorially attached political community that is formally organized around the principle of residency, with boundaries, distinctions between members and non-members, and democratically elected officials, who are expected to pursue policies that benefit the members of the community that elect them. We thank Lior Glick for this reference.

[4] Weinstock (2011) claims that a city is characterized by a certain degree of spatial integration.

distinctive character, relying on some form of local, urban citizenship, which makes an impact on how city-zens think politically (Bauböck 2003; 2019). The distinctive local character possibly helps us to divide what may look like an undifferentiated conurbation into a number of separate cities.

Fourth, as Daniel Weinstock (2014; 2011) notes, perhaps following Aristotle, the city is the smallest geo-political unit in which a person can find anything they need and want to do.

On the basis of these four factors, it is clear that cities are significant political bodies with vital decisions to make about service provision and strategic planning. Although city-level decision-making may often be presented as a largely technical matter, it will always be informed by values, whether or not those values are made explicit. Contemporary metropolitan cities often pursue normative goals such as explicitly aiming to be a low waste city, a smart city, a resilient city, and so forth. Indeed, when we consider the well-being of individuals, what happens in their city can be just as important, or perhaps even more important, than what happens in their nation. Of course, only a nation can declare war, or make fundamental changes to a tax code. But a city can pursue a strategy for green spaces, for public transport, for street cleaning, and so on, and these affect people on a day-to-day basis to a very high degree. Furthermore, a city has to decide how to allocate its budget to different districts, either directly, or indirectly through the location of services such as fire stations or libraries. Some countries, to some degree, follow the 'principle of subsidiarity' which proposes that decisions should be taken at the lowest level consistent with rational efficiency, thereby combining efficiency and local autonomy, although in other cases power is jealously hoarded at the centre, as far as possible. But the general trend has been to disperse power, or at least responsibility, and as a result cities are becoming politically and economically more important and independent than they were. Many services are now the responsibility of local government (housing, police, education, social work services). So there are many acts of decision-making which make an impact, and therefore should be guided by values and moral considerations.

At this point we should add that for the purpose of this book our concern is *metropolitan cities*, by which we mean major cities that are politically, economically, and culturally significant and serve as centres for larger populations than the city's residents. While metropolitan cities are not necessarily huge in their population or size (i.e. they are not necessarily mega-cities) they often have a densely populated urban core and a less-populated surrounding area, often comprising suburbs or small towns which can be politically independent, but nevertheless rely on the metropolitan city for important services such as hospitals, colleges, markets, ports, or airports. But we also add that for a city to be considered metropolitan it has to think of itself as a political alternative (at least in some spheres) to the state; namely it challenges state regulations and policies and offers alternatives in various spheres of life. We can, therefore, regard such cities as 'cities with an ego'.[5] Remember that we opened this chapter by asserting that cities and states are different institutions. Indeed, cities, we want to argue, are not only different institutions from states but also generate different states of mind. When the city-zen, the individual member of a city, reflects upon herself and the state (as a citizen) she has in mind different expectations, images, and metaphors compared to those she associates with when she contemplates about herself and the city.

As Iris Marion Young suggests, the modern city has an important mediating function. It can avoid, on the one hand, a suffocating tendency to allow the community (or the state) too much power, determining and controlling each person's behaviour, and, on the other hand, an alienating individualist tendency, and complete privatization of our social life (Young 1990). The city allows us to smoothly switch from one mood or state of mind to another. In the morning we can go to work and be individuals who seek to flourish and promote our particular interests (including caring for those to whom we have special obligations, such as our children, by earning a salary), and in the afternoon or evening we can become part of the city, participating in activities in our neighbourhoods, or clubs to which we belong, or simply going to a pub or restaurant where we relax alongside others. Even when we escort our children to the playground, on the way back home we might pop into the local farmers' market, and in the evening, when we visit the cinema, our state of mind is different and becomes more communitarian in all these acts. We see the well-being of our neighbours, or of the institution of the farmers' market, as part of what constitutes our own well-being, our identities, and even contributes to our joy.

[5] There are many examples of cities that challenge the state nowadays, and in some countries this happens more often than in others. Here are four examples: First, consider a decision by NYC's municipality to ban selling cigarettes to those under 21 years old, whereas in the rest of the country the law bans cigarettes sales to those under 18 years old. Unfortunately (if we may say so) for those who smoke, smoking is important. Such an act is meaningful to them. Or consider Berlin's reaction to the German federal court's verdict that circumcision was illegal. Berlin declared that it would allow circumcision. Berlin, of course, is also a state within the German federation, which is why it could do so, but those living in the city interpreted it as a challenge to the federal state by the city. Third, in Israel a law prohibits shops from opening on Saturday. Enforcing the law is in the responsibility of local authorities. Tel Aviv municipality decided it would not enforce the law. Fourth, in the United States, especially when Mr. Trump was president, dozens of cities declared themselves as sanctuary cities, asserting that they would not cooperate with the federal government in enforcing immigration laws.

1.3. The Egalitarian Spirit

In this book we ask what it is for a city to embody the egalitarian spirit. This is a precursor to the more practical question of the nature of the policies a city should pursue in order to become a city of equals, though what we say is also designed to inspire policies (and also warn against inegalitarian policies) as we will briefly explore in the final chapter. Aiming at equality is likely, at best, to be just one of several ideals that a city follows, and we do not argue that a city should pursue equality to the exclusion of everything else, such as environmental goals or understanding its special responsibilities in relation to other regions of the country, although of course other goals can sometimes reinforce equality rather than detract from it. But we want to know what it would be to embody the egalitarian spirit. To put it another way, if you are an egalitarian, what should you wish for in a city?

Many people around the world are concerned about inequalities. Some worry because they care about inequality itself; others, such as Frankfurt (Frankfurt 1987; 2016), because they want to ensure that the least advantaged enjoy 'sufficient' (whatever that stands for) resources, or welfare; and a further group, broadly following Rawls, believes we are under a moral obligation to provide for the least advantaged first, as a matter of priority (Parfit 1997; Crisp 2003; Arneson 2013). In the latter case, when goods can be distributed either to those who are well-off or to those whose lives are not going that well, we should prefer the latter as a matter of priority, even if it is *not* the case that the least advantaged can gain more utility from this good.

Each of these ideas—equality, sufficiency, and priority—undoubtably has appeal. First, inequalities can seem unfair, even among the affluent. Second, if some people are living with an insufficient amount to support an adequate life then we can feel something has gone badly wrong in society. And if there is economic growth but those at the bottom do not share in the success then we feel that a society has lost its sense of justice. There is, of course, sharp philosophical debate about which of these ideas should take precedence when they conflict. However, as we argued in our previous book *Disadvantage* (Wolff and de-Shalit 2007, 2–4) in circumstances of scarcity and in which there are still people who have not achieved sufficiency, all egalitarian principles—equality, sufficiency, priority—converge on the same general policy: egalitarian cities have to identify the worst off and take steps, directly or indirectly, that will improve their position.

Yet as we also argued in *Disadvantage* (Wolff and de-Shalit 2007) there is no simple answer to what it means to be worst off. Well-being is plural, and, we argued, the categories are at least in part not fully comparable or compensable. For the purposes of our previous study we adopted a modified version of the capability approach, following Sen and especially Nussbaum, arguing that what matters from an egalitarian perspective is not merely whether functionings and capabilities are distributed fairly, but also how secure people's functionings are, and whether people have 'genuine opportunities for secure functionings'. In this present work, while continuing to recognize complexity, the change of scale and focus of the study means we will take a far more contextual and less abstract approach, and ask first how we should understand 'being worst off' or treated as an unequal *in a city*? What does it mean to have less, or to be treated as an unequal in the city? Less of what? Unequal in what respect?

This is not an easy question. Imagine you are entering a period in your life when you wish to settle down and perhaps raise a family. You are considering which city to move to. Presumably you will consider the cost of living and how good the salaries are in that city; but you will also think of the local education system, the crime rate, the level of pollution, whether there are nice parks, cool pubs, good theatres, a good enough variety of cinemas, maybe the café culture of that city, its public transportation system, whether you will need a car and where you might park it, how people of your ethnicity are treated, and so on. Thus, we can hypothesize that when people think about what matters in terms of the quality of their lives, and hence equality, in cities they look at their lives at a level of detail which is distinguishable from what they think is important when they think of distribution on the level of the state. This provides a rich basis for thinking about inequality in the context of the city.

However, the pluralism of well-being may seem to make the question of what it is to be worst off in the city intractable. How do we compare someone who does well on one criterion and badly on another with someone who has the opposite profile? How, in other words, do we weigh the different parameters? In *Disadvantage* (Wolff and de-Shalit 2007) we suggested that this is more of a theoretical question than a practical one, for disadvantage tends to cluster in the sense that people who do badly in one respect often do badly in others, and there are causal mechanisms that explain why. Here we do not abandon that framing of the issue, but we add a further element that makes 'all things considered' judgements considerably less important. We believe that being treated as an unequal could consist of doing especially badly on one or two parameters, such as being a victim of racism, or being denied civic services available to others in a similar position, rather than an all-things-considered judgement of total well-being. Hence if, as sometimes reported of some cities, wealthy members of racial minorities find that taxi cabs do not stop for them, we do not say 'your wealth makes up for the discrimination' but rather that the city fails to meet one of the obvious criteria for embodying the egalitarian spirit.

In that previous study we attempted to develop a position that was sensitive both to questions of distribution—the typical terrain on which the 'equality, sufficiency, priority' debate takes place—and to questions of social or relational equality, or, in other words, the question of what it is to relate to each other as equals. When we, the authors, consider the question of what attracts us to the idea of equality, we find we are less interested in making sure that everyone has the same amount of anything that can be distributed between them, but rather that each person has good reason to regard each other as an equal, and be regarded as an equal by them. In the words of R. H. Tawney, for us the enemies of equality include snobbery and servility (Tawney 1931), though with Iris Marion Young we would add exploitation, powerlessness, marginalization, cultural imperialism, and violence (Young 1990) and also include social exclusion too, as well as, perhaps, other relations (Wolff 2017; 2019a). For this reason we find Nussbaum's capability of 'affiliation' especially important, as it can capture all these parameters. This is especially true for affiliation in the city, which, as we have defined it, is about connectedness. So in a way this book can be considered as a study in what a secure sense of affiliation means at the level of the city.

From these considerations it may already be apparent that we are less interested in precise measurement and comparison than many of those writing about equality in the city have been. In this respect we are influenced by the joke with which Harry Frankfurt starts his famous paper 'Equality as a Moral Ideal' (Frankfurt 1987).

FIRST MAN: How are your children?
SECOND MAN: Compared to what?

Frankfurt's point is that there is something alienating and disconcerting about making comparisons. It seems competitive in a way that it is contrary to the considerations supposedly motivating egalitarians. But this point needs to be considered carefully. Frankfurt does not suggest that we should never compare. I can compare my life with my successful neighbour in order to discover what my life lacks, just as I might have found out what is missing in my life by reading a book or watching a movie. And indeed, for some goods strict equality will be necessary, such as in the distribution of votes in local elections. But generally speaking, we are interested in whether people are able (rightly) to regard themselves as being taken to be an equal in their city, and in how to translate this into a kind of moral principle, or a principle for policy, rather than counting how much of each good they have compared to others, even if there are times when counting is entirely appropriate.

Consequently, in this book, somewhat paradoxically, we suggest that a sense that you are being treated as an equal in the city cannot be entirely reduced to the claim that there is a scale on which you are being measured and come out as equal, and especially we deny that equality in a city can be measured by a comparison of possessions or resources. Nevertheless, undoubtably there are elements that can and should be counted and compared. We would be concerned to find that garbage collection happens more regularly in wealthy parts of town than poorer parts, or having an address in a certain neighbourhood makes you less likely to be called for job interviews for example, or you might even be unlikely to reach the interview on time because of very poor public transport services (Giannotti and Logiodice 2023). But at the same time there are intangibles that are much less amenable to precise measurement, such as a feeling of being respected by the city authorities, although even here surveys can act as a proxy. Hence there could be a complex set of indicators that would allow us to judge where particular work is needed to make a city more of a city of equals, and we will return to this in the final chapter. Nevertheless, we would be alarmed to find that a city has and uses such a scale as anything more than a rough heuristic to guide policy. Detailed attention to how people perform on the equality scale could mean the city has made a fetish of equality rather than seeing it as part of an organic, intrinsic, element of city life.

1.4. Methods

The city population's fluidity raises another important methodological issue: inequality among whom? Who should be counted as the city-zens? Obviously, residents in the city should count. But under this category we would also include immigrants before naturalization—individuals who often are not counted when inequality at state level is measured. No less challenging is the question of whether to include commuters (individuals who work and shop in the city but do not reside there) as well as non-resident tax payers (individuals who own a business in the city but live elsewhere). Voting in local elections, for example, is possible for some commuters in some cities (e.g. the City of London)[6] and for non-resident tax payers (e.g. in

[6] This is an anomalous case as what is known as the City of London is a small area in the centre of London, with few residents, and many businesses. Some non-resident business owners can vote in local elections, as well as residents.

Australia).[7] Such people have an interest in what goes on in the city and contribute significantly to its economy, so prima facie the demand to refer to them when measuring inequality in the city sounds reasonable. Yet it is also true that there are also people who never enter the city but have important relations with it, and a legitimate interest in how it is run. For example, the headquarters of the supermarket they shop at, or the bank they use, could be located there. It is hard to draw a natural line about whom to include and whom not to. We have chosen to focus on those who live within the city boundaries, because they are those mostly affected by the city's policies, and are constantly subject to the city's regulations, but we are conscious that other choices could have been made.

We have already emphasized that inequality, the personal sense of inequality, and the study of inequality in the city, differ significantly from the same issues at the state level. First, individuals belong to various institutions with different norms or ethical principles, and their political consciousness, states of mind, and expectations change when they switch from thinking as citizens of states to thinking as city-zens of cities (Magnusson 2011; Amin and Thrift 2017; Bell and de-Shalit 2011; de-Shalit 2018). Löw (2013) argues that city dwellers regard their cities as 'entities of meaning', expressed through the quite different types of attachment they feel towards their cities, in contrast to their countries. Compare the open and liberal slogan 'I love Berlin' with the chauvinistic overtones of 'I love Germany'. The former sounds celebratory and welcoming whereas the latter reminds us of fearful moments in history, or in any case has nationalistic overtones. This might suggest that people feel differently towards the city and the state, and develop very different expectations from them, and that we as society regard such feelings differently.

Second, there are likewise different states of mind concerning equality: at state level, for many people, income and wealth; at city level, what we have access to—the education system, levels and distribution of crime and pollution, pleasant parks, cool pubs, good theatres, café culture, housing, public transport—as well as, critically, how we are treated. As we will demonstrate later in this book, a city's qualities, amenities, services, and social relations are critically important to its egalitarian character.

Third, in thinking about a city of equals, we must particularly attend to the dimensions that local authorities can influence. As much as personal income and wealth matter, local authorities can do very little about them

[7] The question whether commuters and non-resident tax payers should vote is discussed in Glick (2021).

directly, especially income. However, as David Harvey has argued (2009 [1973]; 1985), state policies affect the value of land and other resources, and thereby have powerful indirect redistributive effects. Harvey's observation is reinforced by the fact that local authorities can mitigate the pernicious effects of inequality through policies such as land-use regulation, zoning (Macedo 2011), development or conservation, or special services (Wolman 2012). They can provide infrastructure (Marsh et al. 2010),[8] attract capital and business, or enable political participation. Indeed it is part of the natural functioning of a city to decide where to allocate its budget and locate services, such as fire services and schools. Hence there is wide discretion, with far reaching consequences for many aspects of the quality of life.

Fourth, another reason why our constant theme of this book is that we do not want to identify equality in the city with its distribution of wealth and income between its residents, is that standard measures of wealth or income inequality can be problematic at city level. While income gaps as measured, for example, by the Gini coefficient index, make sense at a national level, they can be highly misleading for the city (Alster 2022). For one thing, the Gini index assumes a stable population. Yet people move in and out of cities, partly in response to policies. Suppose a city makes life difficult for poor people, manipulates them into leaving, and busses in labour from outlying districts, as reported of Giuliani's NYC (1994–2001) (Baker 2005; Polner 2005). Such policies intuitively feel anti-egalitarian, but their immediate effect is to lower the Gini coefficient.

But even if these technical issues around inward and outward flow of population can be overcome, Gini measures, and similar, are simply a poor match for our intuitions about which cities best embody the egalitarian spirit. Consider once again liberal Berkeley, which we have already mentioned as an inspiration for this study. Berkeley hosts homeless people as well as many students, who are classified as low-income regardless of parental income. Paradoxically, in part because of their liberal and egalitarian policies, in Gini terms Berkeley is exceptionally unequal (Knobel 2014). This alone is enough to make us look for alternative accounts of equality in the city. To recycle the motivation of the capability approach, what matters is less what you have, but rather what you can do and be. And this will depend on what the city offers, in terms of infrastructure and public services, among other things, alongside the market. Thus, rather than income gaps, we argue that what matters in egalitarian terms is that city dwellers are able to build valuable lives for themselves

[8] For example, upon becoming Commissioner of the NYC Department of Transportation in 2007, Jannette Sadik-Khan realized that the city lacked resting spots in public spaces when she saw people perching on fire hydrants (Sadik-Khan n.d.).

and have a secure sense of place, independently of their economic success. This idea—that everyone has a secure sense of place—is the core for us, of what it is for a city to embody the egalitarian spirit.

The two questions that jump out in the face of such a claim are, first, what does the idea of a 'secure sense of place for all' mean in detail, and second, how do we show that this does indeed capture what we are calling 'the egalitarian spirit'? Answering the first question is the project of this book. On the second—the issue of justification—we cannot claim to offer any sort of demonstrative argument, although the empirical grounding that we discuss in Chapter 3 and 4 shows that the account we draw resonates with the views of many city dwellers. But instead, we try to draw up a picture that will be as compelling to others as it is to us. We hope that those with egalitarian leanings, on reading our account, will think, 'that's the type of city I want to live in'. Although we are painting a general picture that is, to some degree, abstracted from particularity, it is important to understand that our account is grounded in particularity, albeit multiple particularities, incorporating not only our perspective as authors, and the views of the academic community represented through a literature review, but also an extensive set of interviews in which we explore with citizens of ten cities what, for them, makes a city equal, or at least feel that it treats them and others as equal.

To explain, between 2015 and 2019 we conducted 182 face-to-face interviews in: Amsterdam (19 interviews) Berlin (20), Hamburg (18), Jerusalem (32), London (13), New York City (11), Oxford (14), Rio de Janeiro (20), Rotterdam (5), and Tel Aviv (30). We use the *dynamic public reflective equilibrium* method that we introduced elsewhere (Wolff and de-Shalit 2007; de-Shalit 2020; Wolff 2020). The first step was to draw on our own reflection and scholarship as theorists, thereby generating a broad conception of what matters in a city from an egalitarian perspective, based both on our own perceptions and arguments, and on an extensive literature review which we will discuss in Chapter 2. We rejected an exclusive focus on economic factors, and incorporated considerations important to egalitarians, such as whether the city starves poorer neighbourhoods of civic amenities, whether all groups have similar access to services, the degree to which groups are, against their will, segregated residentially, and how individuals feel they are treated.

The next stage was to test our understanding with residents of cities, as they have distinct knowledge and important normative intuitions about this question. This was the point of our interviews with ordinary, randomly selected city-zens of differing ages (17–85), genders, sexual orientations, social backgrounds, and ethnicities. We use the term 'interviews' because it is used in social sciences, but in some respects it would really be more accurate to describe them as philosophical conversations. We acknowledge that some social scientists, who are used to structured interviews where the interviewer does not engage in discussions and debates with the interviewee might find this term, interview, misleading, but no term is perfect.

We interviewed, in the sense of having conversations with, long-standing residents and newcomers, sampling numerous districts within each city, often at different times of day. Our overall goal was to enrich, deepen, and develop our understanding, and to inspire us to insights we may not have reached had we relied purely on conventional academic resources. In short, we wanted first to see if the interviewees confirmed, or conversely challenged, the aspects we had already identified as important to the understanding of a city of equals; second, to fill out our broad conception with richer detail; third, to see if we had overlooked any dimensions; and fourth, more broadly, to provide inspiration for our own thinking. To summarize, our goal was not to provide a statistically significant survey of views, but to enrich our own understanding. Hence we could call it a methodology of 'enrichment'.

We like to think of our interviews using the metaphor of springboards, for which we thank one of the book's referees. Although the notion of a 'springboard' has become something of a cliched metaphor to mean something like 'assisted starting point' we would like to revitalize it by taking it as something more like a trampoline. Jumping from the springboard or trampoline enables you to gain an elevated viewpoint, thus freeing yourself from the solid ground of your position. Similarly, we wish to challenge our initial theory, look it over, and be able to revise it if necessary. Also, jumping so high, we can see things we did not see before when standing on the ground. Similarly, these interviews enable us to listen to questions, ideas, and thoughts that are lacking in the literature. These semi-structured interviews—perhaps better described as barely structured at all—of between twenty and forty minutes, followed a pre-prepared set of questions, but allowed plenty of space for the interviewees' own reflections, including us challenging their views by applying analytical philosophical methods, as we would do in class with our students. They started with the interviewee reflecting on what is important to them in their urban experience (to inform relevant dimensions of inequality), and moved to questions about inequality in their own city. Finally, they were asked to justify these normative evaluations by proposing an ideal of an egalitarian city and corresponding policies, asking such questions as what they would do if they were 'mayor for the day'. We often challenged the interviewees' normative standpoints, asking them to propose a rationale.

Although it was important to consult with and be inspired by city dwellers, we did not cede absolute authority to the interviews. This is because we did not aim at gathering data to be later analysed. Nor did we wish to reach a representation of what the public thinks. Rather, as noted, the point was to inform, enrich, and modify our own, necessarily partial, perspectives. Therefore, the range of responses, rather than the preponderance of particular answers, matters most. For similar reasons we also consulted surveys such as 'Soul of the Community' (Knight's Foundation and Gallup, n.d.) and the Eurobarometer 'Quality of Life in European Cities'. While the focus of these surveys is not inequality in cities, they explore what matters to city-zens, thereby providing materials to inform our own analysis concerning dimensions of inequality.

Because our method is analytical political philosophy and yet we rely on interviews and qualitative research, we realize that we are open to various lines of question. One possible criticism of our interviews is that the sample is too small to be statistically significant, and furthermore we did not code and analyse the texts using a formal methodology. But this would be to misunderstand the purpose of the interviews. We did not attempt a survey to provide an empirical, authoritative, account of 'what city-zens think'. The function of these interviews is not to find out, empirically and statistically, the preponderant views of the city-zens in any city, but to serve as an inspiration for us when we philosophize and suggest conceptual and normative arguments. These interviews are for us what texts of other philosophers are for many of our colleagues: a point of departure, an inspiration.

Furthermore, a cross-city study such as this one will face the difficulty of cultural attitudes to expressing oneself to outsiders. In some cities interviewees may want to defend their cities from outside criticism, and hence over-praise it, whereas in others interviewees may be openly, or even overly critical. This type of variation is, however, not a problem, but a valuable resource for us, as we are not attempting to collate and compare results. Rather, as said, it was to inspire, challenge, illustrate, deepen, and inform, and so what matters is that views are expressed, not how many people said them. Moreover, it is important that we discuss these views with the interviewees, because in this process we and they sharpen our thoughts about the issues discussed.

Another possible challenge is that in this book our understanding of a city of equals will be heavily impacted by the choice of cities in which the interviews were conducted. For example, it might be that we reach a notion of a city of equals as one that embraces and integrates immigrants simply because we chose cities that already had this attitude. We disagree with this challenge. First, we did conduct interviews in cities where immigration is often welcomed, most notably Amsterdam, but also where, arguably, it is much less so, such as Rotterdam and Tel Aviv. Second, and most interestingly, while there is some variety in what city dwellers in the various cities mentioned as the components of an egalitarian city, in general, nearly all the topics were mentioned time and again by city dwellers from all the cities.

Nevertheless, to be on the safe side, we conducted the interviews in different cities in several countries, where these cities differ by size, their claim to fame (whether a capital city, a commercial centre, or a cultural centre, etc.), their dominant religion, whether they are in a liberal or less-liberal country, and many more variables. And still, the parameters repeated themselves in those cities. Third, and most importantly, we emphasize again, that we did not aim at reaching a representative picture of what city dwellers think. Instead, we wanted to capture the insights of our interviewees in order to challenge ourselves and be inspired.

But we acknowledge that the methodology of stopping people in the street has its own biases.[9] Those who regard their time as critically important are less likely to stop, and so some groups, such as people who are retired or not in the formal workplace, are more likely to be represented than others. We attempted to compensate by seeking out as much variety as we could, by conducting the interviews in different parts of the cities and in different times of the day, and among veterans and newcomers, men and women, the young and the elderly, people of different ethnicities, and so on. We also used research assistants who spoke the local language in German and Brazilian cities, as we do not speak either German or Portuguese. Having said that we are aware that we could also have missed some voices. But at the same time, as we are not pretending to offer a representative view, this criticism is not as significant as it could have been for other types of study.

Even though we do not claim to have a value-free starting point, there could be a lingering concern that the study suffers from confirmation bias, especially as our interview script was driven by our research agenda, including our provisional suggestions. For example, we were not interested in the empirical question of how many people believe that the city should be egalitarian, but more in the question of what it means for a city to be egalitarian. Right from the beginning, therefore, in effect we excluded many opinions from the interviews. But this, again, is legitimate, as we were not interested in whether people are egalitarians or not, but in what an egalitarian account of the city

[9] A growing number of researchers believe that one cannot write about urban issues without strolling, and being inspired by talks with people, as well as interpreting the city's planning, architecture, and design. See, for example, Sharon Meagher (2007).

would entail. So to conclude, without entering into the debate about whether it is possible for any study to be entirely value-free, we accept that our own starting point was value-laden. The remedy, such as it is, is to be aware of these possible biases, and to be especially vigilant in looking out for ideas that conflict with our own. And we did find such cases. For example, one Berliner reported that they felt excluded in some social contexts because they were 'too conventional for bohemian Berlin', a very different way of looking at exclusion in the city than our initial understanding of it, and we were challenged by a small number of interviewees who stated views that could be interpreted as racist, especially when complaining about the way in which their neighbourhoods had changed over the years. In one case (in Jerusalem) an interview was stopped because the interviewee expressed views which were so xenophobic and anti-Arab that we realized there was no way these views would be interesting and relevant for our effort to understand what the egalitarian city is.

1.5. The Argument in a Nutshell

To summarize what we have said, and where we are going next, one obvious answer to our question, what makes a city a 'city of equals', is income and wealth equality. But we see that some cities, although they have unequal distribution, are nevertheless regarded as 'egalitarian' in spirit, which is a puzzle and one of the main drives for this research and book. Starting with our own reflections, we discussed this question with city dwellers in ten cities in six countries and three continents. This helped us refine our own view and construct a theory. These interviews aimed to challenge and enhance our opinions and inspire us.

Reflection on our initial thoughts, the literature review, and the interviews enables us to come to a structured view of what we think it means for a city to offer each individual a secure sense of place, and thereby be a city of equals. In such a city, people feel, and are treated as if, they have as much right to be there as anyone else; there are no off-limits spaces in the city (apart from exceptional cases such as parks or halls designed for women only), city dwellers feel they are part of the city's story and have pride in the city. Critically they also feel that the city has pride in (people like) them, rather than wishing they were not there or ignoring them.

More formally, we argue that a secure sense of place consists of four core values: (i) access to the city's services is not constituted by the market; (ii) equal opportunity to achieve a sense of meaningful life; (iii) diversity and social mixing, without a monolithic culture; and (iv) inclusion without deference or submissiveness (by which we mean that city dwellers should have access to the facilities and resources of their city by way of automatically assumed entitlement, rather than as grudgingly granted discretion, relying on the mercy or discretion of gatekeepers).

A question might arise about the four core values. Isn't it the case that sometimes there must be some trade-off between these goods? We will explore this in more detail in Chapter 5, but here we would like to clarify that we do not necessarily believe that all good things go together and that life can be perfect. Some things come into conflict with others. For example, we were impressed to see that in Rio de Janeiro the city operates public football pitches that are open throughout the night. When we raised this as an example of treating people as equals because it caters to the needs of those who work at night, often people working in hospitality and transport, we were told by locals that, well, many who reside in the favelas cannot make use of the facilities because public transportation at night is awful. We admit that this creates a kind of 'elite' among the working class, or that it discriminates against those who live far away, and that ideally, public transportation should work during the night, or, alternatively, football pitches should be arranged in additional areas of the city. Yet, the city is more egalitarian if it enables some working-class people to play at night. Trade-offs are a fact of life, and city planners will have to face them and find reasonable solutions. Moreover, cities vary, even more than states. Consider Glasgow (Victorian, big-city vibe, fun city, great nightlife) versus Edinburgh (more medieval, traditional Scottish, more bourgeois, pretty city), or Milan versus Rome, etc. These cities might have a different ethos from each other, with different values and priorities, which to us adds to the richness of life. Ideally, people can then choose where to live and which city suits them best. Of course, in real life, there are many considerations about where to live, and many people do not have the good fortune to live in the type of city that would be ideal for them. Nevertheless, it is important for us that everyone can live a meaningful life wherever they live, and accommodating such a value should be a high priority for all cities and perhaps a way to navigate some of the inevitable trade-offs between different values. But given that circumstances vary so much we are very reluctant to give any value guaranteed priority over others.

Our findings generate a pluralistic account of the egalitarian city, although we do not attempt a strict definition. Instead, we offer a 'family resemblance' account (Wittgenstein 1953) of the features that in varying combinations can make a city egalitarian. We will set out and develop our own findings in Chapters 3 to 6.

1.6. Conclusion

Our task in this introductory chapter has been largely explanatory; to explain why we are interested in the question of equality in the city, how it may differ from equality in the state, and to lay out the outlines of the methodology we adopted in pursuit of our theory of what makes a city a city of equals. We have also given a very brief account of the definition we adopt.

In Chapter 2 we undertake a literature review. Had we restricted ourselves to political philosophers who have directly discussed equality in the city it would be rather short. Since the book is written for a wider readership, including social scientists, urban studies scholars, planners, and political scientists, we have extended the discussion in several dimensions. Thus, we first look to contributions from a broader range of social scientists, including urban studies and sociology, as well as political philosophy. Second, we incorporate a broader discourse on justice in the city, because justice is not often explicitly distinguished from equality in this literature. And third, we look at some contributions which indirectly bear on our questions, even if they do not address them directly. Chapters 3 and 4 summarize our interviews, drawing out a series of key, recurring, themes. Once again, these chapters might appeal more to social scientists than to philosophers. Those who wish to skip them, or read the theory first, can jump to Chapter 5. It explains our theory in detail, interweaving observations and results from the previous chapters, while the final chapter, Chapter 6, explains our approach towards operationalizing our theory and makes some initial steps and suggestions.

2
A Critical Literature Review

2.1. Introduction

The topic of justice and the city has been widely discussed in urban studies, geography, and economics, but despite the publication of a number of interesting works, it has not yet generated a focused literature in political philosophy. We subscribe to the view that political philosophers interested in the city should carefully, even if critically, read the works of social scientists, especially when they explicitly grapple with questions of justice. In this chapter we will provide a critical review of those works that we regard as most influential or insightful in helping us think about questions of equality in the city. Although few directly address our topic, and many are limited in scope or focus, or put forward positions we may dispute, we have profited from all the books and papers we will discuss. In this introduction we will provide a brief introduction to several of the works we will return to, in order to set out some of the context of the various contributions.

Probably the natural starting point and classic source for any discussion of political philosophy in relation to the city, is geographer David Harvey's *Social Justice and the City* (2009 [1973]). Harvey, very well known for his expositions and defence of Marx, takes a type of dual perspective in this work, starting with a relatively liberal discourse on justice, and then moving to a more Marxist analysis. The second edition of his book, published in 2009, also includes a useful essay on 'The Right to the City'. Harvey is especially interested both in the use of space in the city, in terms of inequality between neighbourhoods, and the ways in which city planners can change the value of assets through decisions around such things as zoning and development. We will return this to in detail below. Another classic discussion of justice in relation to the city is urban planning academic Susan Fainstein's *The Just City* (2010), which takes case studies of three major planning developments in different cities—New York, London, and Amsterdam—and evaluates them in terms of the norms of democracy, diversity, and equity. Perhaps, though, the founding volume of the discipline is Jane Jacobs's *The Death and Life of Great American Cities* (1961), which we have already mentioned as an influence on

City of Equals. Jonathan Wolff and Avner de-Shalit, Oxford University Press. © Jonathan Wolff and Avner de-Shalit (2023).
DOI: 10.1093/oso/9780198894735.003.0002

our own work. Jacobs asks what makes some parts of cities feel safer than others, and her key idea, which we will return to, is diversity of uses of those streets, in stark contrast to the rigid zoning which was becoming fashionable as she was writing. In Jacobs's work concerns of justice are ever-present if not, explicitly, theorized.

We have said that the philosophical literature on our topic is not extensive, but we do not want to diminish or sideline the important discussions over recent decades that are finally generating a much wider set of debates. Significant contributions include Owen Fiss et al.'s *A Way Out* (2003), Elizabeth Anderson's *The Imperative of Integration* (2010) and Tommie Shelby's *Dark Ghettos* (2016), all of which provide contrasting perspectives on a serious injustice in the United States: the existence of largely African-American, deprived, inner-city ghettos in many large cities. We will return to this pressing issue later. Clarissa Hayward and Todd Swanstrom's *Justice and the American Metropolis* (2011a) is an important collection of relevant essays because the papers together address what today is often called the intersectionality of injustice, namely the various aspects of injustice—race, gender, age, segregation, physical abilities, and so on—that often coincide. The book focuses on these aspects and perhaps less on class matters because the editors wanted to shed light on power relations which are not always visible, at least not at first sight. The editors refer to these structural injustices which are not always noticed as 'thick injustice' (2011b, 4). In addition, the volume is vital in raising the question of how to solve these injustices, though, like many urban scholars, the first tendency is to find solutions in urban planning.

While Hayward and Swanstrom do include a section about 'justice and institutions' (in particular, voting), the connection between solving inequality in the city and democratic activism is much more developed in Margaret Kohn's *The Death and Life of the Urban Commonwealth* (2016), which also explores the relationship between equality and democracy in the city. Kohn begins with a very basic observation, which is not discussed widely enough in other works: namely that the city belongs to all of us, all city dwellers, as well as visitors, but many of its parts are not accessible to some, either for economic reasons (the cost of rent or of goods in local shops) or because they feel uneasy in the area. She argues that urbanites lose access to the urban commons, which, we accept, is unjust. Interestingly, Kohn, like many others, focuses on economic or racial barriers, but our interviews, which we discuss in Chapter 3 and 4, indicate that there are other, cultural barriers. For example, we mentioned in Chapter 1 an interviewee in Berlin who said that she is not the type of person who likes fashion and knows how to dress as a 'hipster', and therefore there are certain areas in the city which she avoids going to because they are too cool, and she would feel unwelcome there. Nevertheless it is true that the many of the examples our interviewees reported to us were based on forms of exclusion rooted in class, race, ethnicity, or religion.

The issues of equality and justice in the city have been tackled by political philosophers also in works that focus on broader questions of political philosophy, but provide very helpful contributions concerning the city. For example, Iris Marion Young includes chapters about the city in her major works *Justice and the Politics of Difference* (1990) and *Inclusion and Democracy* (2000), although it is not the main topic of either book. For us, Young's work contains much of great insight and we will return to her in detail several times. There is also increasing attention paid inside and outside the philosophical literature to gentrification, which is broadly described as a particular process of demographic change in a neighbourhood, which is sometimes (but not always) also a racial and ethnic change.[1] Gentrification typically begins when a street or small neighbourhood becomes more attractive to people with higher incomes, who move in, thereby changing the neighbourhood's character, and making it much more difficult for those previously living there not simply because of the change in character, but because rising rent or house prices make it unaffordable. Perhaps the most disturbing and controversial kind of gentrification is where social housing is transformed and privatized. Older buildings, including blocks of apartments are demolished and replaced, or refurbished in whole or part, to be made suitable on the open market for mixed-income communities, but can soon become dominated by the better off. Loretta Lees et al. (2008; Imrie, Lees, and Raco 2009) claim that processes of gentrification, in particular the privatization of social housing, are not inevitable and that not only are there alternatives, such as community self-build or community land-trust, that can cater for the needs of everybody, but that these alternatives are more socially and economically sustainable.

A classic philosophical work on gentrification is Margeret Kohn's 'What Is Wrong with Gentrification?' (2013) where she raises the question of whether individuals who sometimes naively search for better accommodation can be thought to be morally responsible for the displacement of low-income residents. Following this question *Gentrifier* (2018), written by anthropologists and sociologists Schlichtman, Patch, and Hill, discusses one of the key issues in this debate, namely whether we should understand gentrification as a process that is created by individuals and their decisions and

[1] The term was coined in 1964 by the sociologist Ruth Glass.

actions (agency) or as a matter of structure, of big urban social dynamics (see also Kaddar 2020). A structural analysis would require viewing micro-level decisions of the agent, the gentrifier, from a macro-level perspective, for example, 'the economy', or 'the city' which then bears primary responsibility. The debate, though, is by no means settled. Marcuse's 'Gentrification, Abandonment and Displacement: Connections, Causes and Policy Responses in New York City' (1985) is perhaps less forgiving and merciful towards the gentrifier. We read Iris Marion Young's classic example of 'Sandy' who faces homelessness through gentrification, as continuing this debate about agency versus structure. Indeed this example is one of the leading illustrations of the idea of structural injustice in her posthumous book *Responsibility for Justice* (2011).

As can be seen, most of these works address injustice in the cities of North America, although Fainstein advances her analysis by studying Amsterdam and London as well as New York, even if her research is based in the urban planning scholarship of North America. We mention this because broadly speaking, there is a tendency among North Americans to analyse inequality and injustice in the city among racial and ethnic groups, whereas among European scholars there is more of a tendency to focus on class, income groups, immigrants, and often gender, as well. For example, Sako Musterd et al. (2017) argue that the essence of segregation in European cities is the separation between poor and rich, and Mehmet Yorukoglu (2002) focuses on income differences in different spaces of the city, arguing that with increasing density, for both production and trade, the key to success becomes 'location, location, location'. This, he argues, is the core of inequality differences across cities of different densities. Danny Dorling, an Oxford-based geographer, has argued that inequality is a problem because culture divides people and makes social mobility impossible (Dorling 2019). As an example, he points to the startling fact that 'the most common way people die under the age of 65 in Oxford is to die homeless', and that it has been this way 'for over two decades now' (Dorling 2022, 14).

As we have mentioned, among the 'classic' and most cited works, many explore injustice in terms of spatial organization of cities, especially looking at planning law and practice. Prominent examples include Marcuse et al. (2009) *Searching for the Just City: Debates in Urban Theory and Practice*, Pavel (2009) *Breakthrough Communities: Sustainability and Justice in the Next American Metropolis*, and Soja (2010) *Seeking Spatial Justice*. However, these works, do not attempt a specific definition of justice for the city, as distinguishable from justice and equality in the state. Instead they either apply political theory written in the context of the state to the context of the city, or concentrate on how to remedy injustice and improve justice in the city, focusing primarily on issues concerning planning, and relying on an intuitive sense of justice.

We will bring out the main themes of these and other works in what follows, focusing on elements that will help us draw up our own account of a city of equals. But we wish to do more than just review the literature here. In fact, we are claiming that while these studies are extremely valuable, none directly asks exactly the question we are interested in: what is it for a city to embody the egalitarian spirit? Accordingly, none provide an answer to that question, although the questions they ask are often very close to ours, and the answers they provide often yield important insights on which we build, as we will explain. This is an important point. Analytical philosophy often advances by applying an antagonistic approach. That is to say that philosophers challenge previous theories and arguments, showing why this or that is not intuitive, or contradicts a well-established theory. They then offer an alternative theory. But we see our project differently. While we do emphasize the difference between our question and questions discussed in the literature so far, and while we do believe that our theory is novel, we do also acknowledge that we have been inspired by many works by philosophers and social scientists, and we do not intend to reinvent the wheel. For illustration we return, once again, to Jane Jacobs (1961) who, as we noted, asked the question of what makes some parts of cities feel safe. In the course of her answer she drew attention to two major, and connected, issues: diversity of use of space, and chance interactions in the street. Consider an environment where people shop in walking distance of where they live, their children go to neighbourhood schools, there is some light industry, with regular deliveries, collections, and other visitors. In local streets one or two grocery shops open early morning and do not close until late into the evening, and children can play in close-by parks and other open spaces. This type of mixed use may seem untidy, and perhaps noisy and chaotic. To a certain cast of mind, zoning to keep busy, messy, and noisy activities out of peaceful residential neighbourhoods may seem more appealing. Yet Jacobs points out that where there is diversity of use there are more people around who get to know each other, at least by sight, and develop an instinct for when something is wrong and keep an eye out for each other, even if they have no direct relationship. People feel safer, as well as included and welcome. Although, as we have said, equality is not her explicit topic, we take a degree of inspiration from her picture and rich descriptive accounts of urban life.

In contrast to Jacobs's attention to the rich patterns of daily interactions, much of the literature from economics focuses on the much more abstract and data-driven study of economic inequality, and that is where we will start in the next section. Urban studies, however, centres more on spatial factors such as residential segregation by race or class, such as issues around exclusion and inclusion, immigration and diversity. That will be the subject of the following section, before we turn to look at work that has taken up questions of justice or equality, or providing models of the 'just city' more explicitly. We will conclude this chapter with a summary of what we take from the literature and the gaps we intend to fill through our interviews and own further reflections.

2.2. Income Inequality: The Importance and Limitations of Material Inequality

As we indicated in Chapter 1, the most obvious starting point for thinking about inequality in the city is to look at the distribution of income and wealth, and some studies, such as Glaeser, Resseger, and Tobio (2008) do exactly this, although they do admit that measuring inequality in such terms yields unintuitive results (we discuss their argument later). Other studies explore alternative economic parameters for inequality,[2] and show staggering, and growing, levels of inequality in particular cities (for example, the London Poverty Profile, Aldridge et al. 2015). Economists debate the causes of such inequalities, and mechanisms for addressing them (see Long et al. 1977; Baum-Snow and Pavan 2013; and Behrens and Robert-Nicoud 2014).

This is important work, but is limited in a number of ways. For example, David Harvey discusses what he calls the 'hidden mechanisms' of redistribution within a city, including the change in value of property as a consequence of local government policies such as the development of a new school or transport facility (Harvey 2009 [1973], 52–3). Others might be more comfortable describing this as changes to wealth, rather than hidden increase in income, but the general point is that material fortunes can change in numerous ways, and they will not all be recorded in official statistics. Indeed, Harvey notes that simply the growth of a city, and therefore increased demand for

[2] Such as the price of sushi rolls, number of eateries per capita, or least upwardly mobile for fast food workers (Miller and Lu 2019), as well as bike lanes, home ownership, and so on. More sensitively, the focus might have been goods which are consumed by a wider portion of the population, for example, the price of street food in different parts of the city, the cost of renting an apartment or a room, or the price of public transportation for those residing far away from the city centre. However, while these reports grade the cities they do not necessarily say anything about inequality unless their consequences are discussed in the context of justice, as Macedo (2011) does.

prime property, pushing up house prices and rental values, can increase inequality more than direct attempts at redistribution reduces it (Harvey 2009 [1973], 54).

The relationships between rising property prices and inequality in the city is discussed also by Stephen Macedo (2011). He notes that higher house prices near better schools not only have (negative) redistributive effects, but show how difficult it is to equalize educational opportunities in a city. This is especially true in the United States, where, unlike other wealthy countries, schools are funded from local taxes rather than more general resources, and so there is a vicious cycle of educational inequality, because wealthy neighbourhoods pour extra resources into schools. As schools gain in reputation, property prices, and hence the tax base and resources available for local schools increase, with no obvious corrective mechanisms and no incentive for the wealthy and powerful to make changes. Another issue that concerns Macedo, like many other commentators, is the differing prices charged for the same goods and services in different neighbourhoods where counter-intuitively the poor often have to pay more, perhaps because both demand and competition between suppliers is less intense in poorer areas and so shopkeepers need high margins on lower sales volumes to pay rent, and shoppers have nowhere else to turn.

For Harvey such inequalities, and especially the regressive effects of policy, are clearly contrary to justice, and in Part 1 of his book *Social Justice and the City*, called 'Liberal Formations' he turns to Runciman and Rawls to present the outlines of a pluralist approach to justice, drawing on considerations of need, contribution, and merit in a weak ordering (Harvey 2009 [1973], 100) to demonstrate the injustice of such rising inequalities (we explain these ideas in a little more detail later). We should also note, however, that hidden redistribution can work to reduce inequalities, if, for example, the city increases its spending on libraries and leisure centres in deprived areas, or issues more permits for small businesses, galleries, cafes, etc., though such facilities could be co-opted by the wealthier, either by travel or through gentrification, pushing poorer people out to under-served areas.

Valuable though this work is, some of the methodological difficulties with relying on income measures, whether direct only, or incorporating indirect measures, are pointed out by Glaeser, Resseger, and Tobio (2009) themselves. They were among the first to argue that inequality within cities is 'quite different than inequality within countries' (2009, 617). They caution that what, by these measures, will count as the most equal societies are those where 'rich live with rich and poor live with poor' (2009, 618). This is an excellent statement of our unease with income measures. As Douglas Rae puts the point:

> Given the historical flow of American urbanization, a low degree of central-city income inequality almost always arises because the high end of the distribution has melted away. This has happened as upper- and middle-income strata depart for the suburbs at high rates, leaving the poor to fend for themselves in the urban core.
>
> (Rae 2011, 105)

Rae argues that some level of inequality in the city is justified, as it indicates that the city is open to various groups, and more importantly, it implies that the better-off subsidize, through their taxes, services for the worst off. Applying Rawls's difference principle, Rae argues that the inequality is justified so long as it benefits the least advantaged. This is possible, he claims, if you consider that the rich pay local taxes, and thereby some services can be provided for the poor that would not be possible if they were not subsidized by the taxes of richer residents.

While we agree with Rae that income equality is likely to be the result either of policies that exclude poorer residents, or the flight of the more wealthy, we would not want to argue that income inequality always leads either to improving material fortunes for the worst off or to inclusive social policies. Our own view is that there is no simple relation between material inequality in the city and the idea of giving everyone a secure sense of place. Much depends on how the city authorities respond to such inequalities, as we will explain in the following chapters.

We want to make three comments about this. First, while it is true that businesses and shops owners do contribute by paying local taxes, when it comes to accommodation the situation is different. It is often the case that those who rent, such as students, or less-affluent families who do not own a flat, pay the local tax for the flat they rent, rather than the landlords. So it is not only the rich who pay the local tax in the city. Second, and more importantly, in the context of the city, perhaps a more just and inclusive way for the wealthy to work for the benefit of the disadvantaged is for the city to enable and encourage investments in urban renewal that will attract the rich to make business investments in the locations where the poor reside, improving the quality of goods and services in under-served areas, without thereby forcing them to leave, in what Levine and Aharon (2022) call 'in place mobility'.

Third, sadly Rae's hopes that the wealth of the rich can materially benefit the poor in unequal cities appears to have limited empirical confirmation. For example, the urban economist Glaeser found that often, especially in recent years, the richer the city is, the higher its inequality. More precisely, he found that the idea that the higher average income in the city is, the less unequal it is, worked only in the 1980s. By the 2000s this has ceased to be the case. Glaeser found that 241 out of 242 metropolitan cities in the United States became more unequal, materially speaking. The main reason for this was the different kinds of skills and human capital that city-zens had, and the very different returns to skills, especially rapidly growing income rewards in certain sectors such as finance and IT (Tonkiss 2015; minute 36:20). Also, according to OECD research (2018) income segregation, that is, the uneven geographic distribution of income groups within a certain area (Reardon and Bishoff 2011), in our case, the city, increases the higher household disposable income is in a city, and the higher a city's GDP. So it seems that attracting rich people to cities and allowing inequality does not automatically result in more money and services flowing to the poor, as Rae suggests.

Unless a utopian transformation of cities is available, the problem that confronts us is how to increase the sense of equality in cities against a background in which they contain a wide of diversity of people and life experience, including in some cases very striking inequalities of income and wealth. Implementing policies that encourage very rich or very poor people to leave will reduce material inequality, but is the opposite of the solution we seek, which is to make all people feel that they belong, on the same terms as everybody else, or as we described it in Chapter 1 and will develop later in Chapter 5, that everyone has a secure sense of place.

At this point we should clarify that we do not regard income and wealth inequality within a city as morally unproblematic, and we accept that there are powerful reasons for opposing gross and growing material inequalities. Economic factors will be central to the account of equality in the city in numerous ways. For example, some cities raise and control their own taxes, and can make the tax rates progressive;[3] if it does so, the city is in that sense tending towards equality, providing that such policies do not encourage the wealthy to leave. Spending within a city's budget also reveals its nature. If a greater amount of money is spent cleaning the streets of the already wealthy then it appears that it prefers to reinforce privilege than aim for equality. But if it invests in shelters for homeless people, or for women who suffer from domestic violence, then it seems more egalitarian. We can also explore what

[3] In Sweden, most people pay only local tax on their annual income, and the tax is progressive. It also varies, and in more affluent localities it can reach 35.15 per cent whereas in less-affluent localities it can reach 29 per cent only (Swedish Institute 2022).

portion of the budget is invested in affordable housing,[4] what conditions the city puts on new property developments, the average size of small flats in the city, whether there is a local tax on empty flats which are not let, whether the city subsidizes daycare, and so on. For example, the London Poverty Profile (Trust for London n.d.) has developed a new, impressive, and interesting list of indicators to compare and rank London's thirty-two boroughs and the City of London, and offers data, borough by borough, about such parameters as: people seen sleeping rough by outreach workers; rent for a one bedroom dwelling as a percentage of gross pay; percentage of 19-year-olds who lack any educational qualification; percentage of people on benefit payments; premature mortality; and infant mortality; and so on (Trust for London n.d.). Such a detailed list is very welcome as it can supply a much broader and deeper picture of inequality across boroughs or neighbourhoods than referring to income only. Admittedly, this is a picture of inequality between different neighbourhoods rather than individuals, but it can also reveal a lot about inequality between inhabitants of these units.

Our question, though, is how such economic and other material factors relate to what we are calling the egalitarian spirit. At a minimum it is fair to say that we believe a city does not embody the egalitarian spirit if it is not concerned about economic equality, if it does not try to ensure that everyone within its boundaries has a sufficiently good life, and if it fails to give priority to the worst off (to repeat the concern with equality, sufficiency, and priority raised in Chapter 1).

We follow here the footsteps of Richard Schragger (2013), who claims that while the conventional wisdom is that cities can do very little to make the city more egalitarian regarding income, taxes, and transfer payments, they can contribute to creating conditions of equality in other ways. Which ways? Schragger suggests that local policies and regulations should be less inspired by consumerist attitudes: what he refers to as 'the dominant competition paradigm (. . .) aimed at attracting and capturing mobile taxpayers'. Instead, he contends, the city should respond to egalitarian attitudes, such as resisting the privatization of public space, and encouraging the development of small, local businesses. If they do so, argues Schragger, cities can still do a great deal to improve the income and wealth of those who are financially less

[4] In April 2021, four-hundred thousand people were living in NYC in affordable housing provided by the NYCHA (New York City Housing Authority). It is the largest public housing authority in the United States. Around a hundred thousand people were living in other public housing facilities. This is, relatively speaking, a high percentage in a city of 8.46 million people. Affordability of housing is considered a serious problem in the United States, as it is in many countries and cities. In the 2020s people living in American cities have been finding housing to be a serious problem, In 2020, 46 per cent of American urban renters spent 30 per cent of their income on housing, and 23 per cent spent 50 per cent (Schaeffer 2022).

well off, and make more people feel welcome and included. Earlier, when we discussed Jacobs's book, we acknowledged that we are inspired by many scholars; this is another example. We develop such insights further, based also on interviews with many city dwellers, in subsequent chapters.

Indeed, we would be especially worried if the city deliberately implemented policies that aim at increasing material inequalities. More realistically cities very often follow policies in pursuit of economic growth that have the unintended but foreseeable consequence of increasing inequality (Fainstain 2001; Harvey 2019). We do not claim that there are never good reasons for introducing such policies, but we do consider increased material inequality a powerful negative factor that needs to be weighed in the balance, especially when other choices can be taken.

Nevertheless, some cities of similar levels of economic inequality seem to do more than others to embody the egalitarian spirit. How so? There are many factors that we will introduce during the course of the subsequent chapters; but if a city experiences the injustice of gross material inequality, and can do little directly except lobbying the national government, it should nevertheless find ways in which people can enjoy the many elements of an urban good life whatever their level of wealth. This means making economic success, and purchasing power, much less central to the ability to live an urban good life and to enjoy the various facilities and services that cities can and do provide, through the provision of various public services and perhaps in other ways. We will return to this in detail, but the main point is that for a city, one egalitarian response to unjust inequality of wealth is simply to make wealth less important, or, in other words, to prevent a situation whereby lack of wealth becomes a risk to a city dweller's secure sense of place.

2.3. Space and Segregation, Exclusion and Inclusion

In Chapter 1 we mentioned that geographers tend to think about justice in the city in spatial terms. But such a focus is not restricted to geographers. In a recent paper that we find very inspiring, and will briefly return to at the end of this chapter van Leeuwen (2020) claims that the whole point of justice in the city is access to human space. Following Honneth, van Leeuwen argues that it should be a space that is structured to meet the demand for recognition, especially of key human features, namely, basic needs, personal autonomy, and social attachments. We prefer the language of 'secure sense of place' to 'recognition', as 'recognition' lacks the immediate connection with a location, and also has been used in political philosophy in particular ways, and so will have connotations for some readers that takes it out of the context of our present discussion. Nevertheless, despite different terminology, there is good deal in common between our approaches, even if, for us, space and its use is only one part—albeit a vitally important part—of a bigger picture. But certainly, accessibility to spaces such as parks, including transportation, commuting, taking children to school, access to cafes, restaurants, food markets, and shops, all of which are components of urban well-being, are central to the picture of equality and inequality in the city that we shall paint. We return to this point in Chapters 3 and 4 where we discuss the interviews we conducted, as these dimensions of equality were mentioned by interviewees time and again. But here we want, first, to look not only at the advantage of this approach for grasping what a city of equals means, but also at some debates among scholars, and at the limitations of this approach, and what is known and what is less known and less clear about the way justice manifests itself through space in the city.

We noted in the previous section that David Harvey's concern with hidden forms of redistribution occurs in Part 1 of his book *Social Justice and the City* (1973) which he calls 'Liberal Formulations'. Part 2, entitled 'Socialist Formulations', turns especially to spatial factors and how the evolution of cities tends to reinforce spatial segregation, with, he suggests, middle-class families moving to the suburbs, abandoning inner-city areas to deprived populations, and often ethnic minorities. Since the 1970s, though, many cities have seen a reversed trend, with wealthier citizens, especially those without young children in the home, moving back into gentrified inner cities, and poor and working families moving out to badly served suburbs, with limited public transport and other facilities. But Harvey's point remains valid. Cities segregate themselves on social, and often, ethnic lines, though how that takes place is dictated by the more affluent groups. Harvey is self-consciously in a long tradition of writers, going back to Engels, and probably well before, exposing demographic patterns within large cities. Charles Booth's famous maps of wealth and poverty in London streets are a further example of this phenomenon at a descriptive level, although Harvey is keen to look for explanations of segregation in terms of the dynamics of property markets. Other theorists have also explored the mechanisms behind segregation. For example, building on a study of socio-economic segregation in twelve European cities, Musterd et al. (2017; see also Musterd 2006) suggest that socio-economic segregation has increased because of four structural factors: social inequalities; globalization and economic restructuring; welfare regimes; and housing systems. A more recent OECD study (2018) suggests that cities with a higher percentage of migrants also display higher levels of segregation for the bottom 20 per cent of income groups, and that the two together result in socio-economic segregation. On the more positive side Johansson and Panican (2016) show how, in the absence of sufficient national interventions, grass-root groups, as well as cities, engage in various poverty relief activities, which has increased inclusion in various European cities.

Although Musterd (2017) is cautious and suggests that 'relatively little is known about the spatial dimensions of rising socioeconomic inequality' in cities, it can be argued that there is evidence that economic inequality and spatial segregation cannot be completely separated. Massey and Denton (1988) measure segregation in terms of evenness of spread across neighbourhoods, and so, for example, if non-EU immigrants tend to reside only in certain neighbourhoods the spread is not even, and therefore segregation is present.

More precisely, geographers, as well as the US Bureau of the Census, argue that there are five parameters which are related to segregation in what they call an 'areal unit'. First, evenness: a minoritized group is not segregated, when in all areal units there is the same relative number of members of the minority and the majority groups as in the city as a whole. Second, exposure: the degree of potential contact, or the possibility of interaction, between minoritized and majority group members within geographic areas of a city. Third, concentration: relative amount of physical space occupied by a minoritized group in the urban environment. Fourth, centralization: the degree to which a group is spatially located near the centre of an urban area. Fifth, clustering: the degree of spatial clustering exhibited by a minoritized group—that is, the extent to which areal units inhabited by minority members adjoin one another, or cluster, in space.

Space is also related to other parameters which are often linked to inequality. On these lines Nijman and Wei (2020) plausibly argue that in cities, processes of segregation and gentrification tend to involve not the highest incomes (certainly not the infamous '1 per cent') but rather gentrifiers of middle and upper-middle incomes displacing people on lower incomes. In a comprehensive study of inequality in Atlanta, Boston, Detroit, and Los Angeles, O'Connor, Tilly, and Bobo (2001) widen the focus on inequality beyond housing to include the labour market. In the North American literature there is a debate whether inequalities in the labour market are predominately the result of latent racism or whether they are structural and spatial, irrespective of racism, that is to say that racism, whether existent or not, does not play a role. O'Connor, Tilly, and Bobo argue that in these four cities—Atlanta, Boston, Detroit, and Los Angeles—even as old heavy industries gave way to new commercial centres, 'longstanding hostilities' and racial and ethnic segregation continue to exert their influence, generating gross inequalities. The authors argue that it is true that job and hiring procedures require skills that are less common in people of racial and ethnic minorities and that some jobs have moved outside the cities to the suburbs; nevertheless unfortunately, some racist discrimination still remained in hiring, causing severe inequality not only in spatial terms but in the labour market as well. One of their case studies concerns Los Angeles, which is also studied by Bobo, Oliver, Johnson, and Valenzuela (2000), once again exploring the links between race, ethnicity, and gender to housing and jobs opportunities. According to this study housing segregation patterns are repeated, and members of ethnic minorities refrain from searching for jobs in areas where they suspect there is racism. Apparently there is some form of collective memory which influences people's behaviour and even informal segregation. Portland is considered the whitest big city in the United States, and this might have to do with 'severe history of racism that, to this day, permeates all systems and institutions, including our neighborhoods, schools, laws, and housing policies' (Habitat for Humanity 2020).

Other studies home in more directly on racism as a manifestation of urban inequality such as Wilson (2012 [1987]), Bullard (2009), and Powell (2009). Powell claims that the mechanism of private property in practice sustains ethnic exclusion in American cities. Robert Sampson and William Julius Wilson (1995) bring violent crime into the analysis, suggesting the thesis of 'racial invariance', which states that 'racial disparities in rates of violent crime ultimately stem from the very different social ecological contexts in which Blacks and Whites reside, and that concentrated disadvantage predicts crime similarly across racial groups' (Sampson, Wilson, and Hanna Katz 2018).

However, a more recent and less US-focused study by the OECD (2018), suggests that what we regard as spatial factors contributes to inequality in the labour market. When a city is already at least partially segregated it is often the case that there is a lack of easily accessible transport connection to enable travel to employment centres by those living in neighbourhoods with a high concentration of people who are poor or minoritized. This is not necessarily the result of explicit racism, but rather of the market: bus and train companies find that such lines and services are less profitable. The research argues that this hinders job opportunities or at least makes it much more difficult for those who are unemployed to find about new jobs and then reach their work place easily (Pritchard, Tomasiello, Giannotti, and Geurs 2019).

Naturally and understandably, in the attempt to provide a type of dispassionate 'scientific' methodology to the study of inequality and its spatial appearances theorists look to measurable, countable, and hence, material factors in defining and explaining inequality. In doing so the studies tend not to capture what for us is a central issue: how the marginalized and segregated feel when they face discrimination and lack of opportunity. In other words, those approaching spatial questions as quantitative social scientists tend not to apply the relational approach to social justice and to inequality, which makes more room for considering how it feels to be treated in particular ways by others. For example, in his very influential book, *The Origins of Urban Crisis* (1996) Thomas Sugrue describes how red-lining—the refusal to give loans to residents of certain neighbourhoods with above-average risks of default, which were typically heavily occupied by people from ethnic minorities and those on low incomes—had the consequence that African Americans who had been living in these neighbourhoods and wanted, following the economic boom in the city after World War II, to improve their homes or buy new homes, could not afford to do so because they could not find mortgages or loans. This created a troubling cycle of poverty and spatial problems, on top of social discrimination, because the job market and the home market are linked, and people could not find better jobs because they lived in these neighbourhoods, and could not leave the neighbourhood because they did not have enough income, nor could they take loans. Sugrue's work is undoubtably important, and these material factors will play a role in our analysis. And yet, in emphasizing measurable indicators he downplays the subjective, emotional aspect of such practices and events, which for us are a crucial aspect of the nature of inequality in the city. Therefore, we also need to bring out in more detail how people are treated, and the way other city-zens in the city relate to them, think about them, exploit them, and more generally behave towards them.

Getting closer, explicitly, to relational equality, the phenomenon of segregation is taken up in detail in two important recent works of philosophy: Elizabeth Anderson (2010), *The Imperative of Integration* and Tommie Shelby's (2016) *Dark Ghettos*. Similar themes and arguments are explored in the earlier book *A Way Out: America's Ghettos and the Legacy of Racism* (Fiss et al. 2003), containing an essay by Owen Fiss, and ten responses, which in many aspects anticipates the debate between Anderson and Shelby. Anderson, like Fiss, drawing on empirical precedents such as the 'Moving to Opportunity' (MTO) programme, argues that disadvantaged African American families should be subsidized to move to better neighbourhoods to improve their life prospects. Two concerns, however, stand out: first, whether there is sufficient absorptive capacity (both attitudinal and physical) within 'good neighbourhoods' to have more than a marginal effect overall; and second, why the responsibility to make such a disruptive change should fall on those who are already disadvantaged.

Taking up this second point in particular, Tommie Shelby points out that where one chooses to live is an important element of freedom of association, and many people will prefer to stay in the communities where they currently live rather than move to white-majority areas. This is echoed in Sundstrom (2013) and Cassiers and Kesteloot (2012), and also in Young (2002, Chapter 6), who suggests several criteria for observing whether residential concentrations of ethnic minorities reflect people's preferences to live next to each other, or a policy or a social atmosphere of exclusion, which we will explore in detail later in this chapter. One of these criteria that we find especially interesting is whether residents 'know' where racial and ethnic minorities are said to be living and if these places carry associations of danger. Young builds on Peter Marcuse's (1997) distinction between three types of residential patterns. *An enclave* is a voluntary clustering of persons according to affinity groups; *a ghetto* is a concentration of some ethnic group, or lower-class people, as a result of formal or informal exclusion and confinement of this group by a dominant group; *a citadel* is the mirror image of a ghetto: it is an exclusive community of class and/or race privilege, restricting others from living there or, in extreme cases, restricting access to the area, as in a gated neighbourhood. An enclave, Marcuse argued, can be a positive and empowering social structure. Some flourishing African-American-dominated neighbourhoods in American cities, and Chinatowns, are good examples. A less-known case is that of the Portuguese-speaking community in South London's South Lambeth Road who have chosen to live together as a matter of self-selection. Many non-Portuguese-speaking people are attracted to visit this neighbourhood because of its uniqueness, or what we might term, its flavour. Indeed, a city that has lots of flavours in that sense is both attractive and egalitarian. We will return to this in detail in Chapter 5.

To paraphrase Shelby's concern in Marcuse's language, for him the onus is to improve local facilities turning ghettos into enclaves rather than to encourage people to abandon their neighbourhoods. But Shelby's main concern in looking at the experience of African Americans living in inner-city ghettos is not so much to attempt to come to an account of equality or inequality, but rather to consider the morally appropriate forms of response to the deep injustice of living as a victim of an unjust basic structure. In the course of his analysis he points out many ways in which African Americans living in the inner city face worse life prospects than their white co-citizens.

As we have already noted, the topic of racial injustice has especially exercised those studying US cities. An excellent example is Hayward and

Swanstrom (2011b) in their introduction to their edited collection *Justice in the American Metropolis*. Their focal point is the 1948 US Supreme Court ruling in Shelley versus Kraemer that states could no longer enforce racial restrictions for sales even if in private law this was not illegal. However, as they argue, the fact that African Americans, in some limited degrees, had the capacity to overcome spatial exclusion did not guarantee overcoming economic inequality and deprivation. They convincingly claim that at least in the United States, political and legal institutional structures contribute to injustice in cities, or what they term 'thick injustice', as these structures make it very difficult to change anything. One reason this is so is the privatization of governance which created fragmented institutional structures. Another reason, they argue, is that these structures make it difficult to comprehend where the injustice lies and what its source is.

This argument echoes a claim made by Iris Marion Young (2002, 207–10), that in many cities the privileged do not realize how privileged they are because of spatial (often racial) exclusion; they don't mix with the disadvantaged and therefore are not aware of their privileges:

> Life does not feel privileged for the white family with two working adults (. . .) Being able to stop off at a gourmet grocery on the way home, to count on police protection and snow removal and to walk or drive a short distance to see a first-run movie seem like the most minimal rewards for an arduous week of work. Segregation thus makes privilege doubly invisible to the privileged, by conveniently keeping the situation of the relatively disadvantaged out of sight.
>
> **(Young 2002, 208)**

2.4. The Importance and Limitations of Spatial Analysis

Spatial issues are important partly because they can be manipulated by policies of zoning. Gerald Frug claims that contemporary American cities are divided into neighbourhoods and zones, to the extent that people from one area feel uncomfortable walking in another area, but, he hastens to add, this inhospitable situation is not the intended result of individual choices about where to live or work; rather it is the consequences of many local government regulations, such as, but not only, zoning (Frug 2001). Still, while spatial issues are central to inequality in the city there is a question of the degree to which they should dominate the analysis of a city of equals. We are not claiming that they do not matter. Take, for example our acknowledgement that factors of inequality of income and wealth can be important. But would they be so important if they were not spatially represented through segregation? And yet, we are arguing that spatial aspects do not reveal the entire picture; hence assuming that inequality in the city can be analysed through spatial analysis alone is misleading. Here are a few examples of why spatial analysis doesn't reveal the whole picture of inequality in the city.

Consider the possible link between segregation and political representation. For example, Cassiers and Kesteloot (2012) argue that on top of the more discussed consequences of segregation—fewer opportunities for income, as one lives far away from the economic centre of the city, and lack of social network and social capital, which in turn make social mobility even harder—there is a third: stigmatization (both of the area or of people) and lack of political representation, in that people living in more deprived areas are often overlooked by decision-makers. But the picture is complex, as, paradoxically, segregation can sometimes assist local political organization. It may help integrate immigrants who look for similarities in culture in their city of destination and may yield common political agendas which in turn make political organization easier.

People have common interests that have to do with their locality and neighbourhood. Now, as Patti Lenard (2013; 2015) argues, recent immigrants (most of whom lack voting rights) tend to be attracted to live next to longer-standing immigrants from their original country, and therefore reside in the same neighbourhood; this implies that if the electoral district, the constituency, is not the entire city, but rather the city is divided to several constituencies based on neighbourhoods, then neighbourhoods where there are many immigrants who have not been naturalized and therefore lack a vote, are not fairly represented.

Lenard's point reveals a possible limitation to an argument made by Loren King (2011), to which we are otherwise sympathetic. King suggests that justice in cities is more procedural than resting only on principles of distribution. He holds that for a city to be egalitarian it must equally respect all its city-zens, including those groups that are often marginalized from the public domain. Thus, in order to see whether a city's regulation or policy is just and egalitarian, we should look whether its rationale appeals to and takes into account the values of all groups in the city, as he defines it, whether it gives equal political standing to all who will be affected by the policy. This is a very appealing suggestion and provides a dimension of equality that takes us beyond pure spatial issues. Still, as Lenard argues, when it comes to political equality there seems to be a need for actual representation of people from all groups and not merely to give them equal weight. The question is how should we give equal weight to the interests of immigrants who are not yet naturalized and therefore lack voting rights. Lenard argues that considering their interests isn't sufficient. This actual representation, as Iris Young insists, guarantees that the narrative of those marginalized will be put forward the way they would like it to be put forward. Representation is of real value, not merely of symbolic value.

However, as the above-mentioned example shows, the motivation for separation is not always negative, and separation of groups in spatial terms is sometimes the result of an egalitarian policy. Practices that enable immigrants to live next to their relatives and friends from their country of origin can be beneficial, and as long as locales don't, in effect, become ghettos, proximately to those of similar origin who have already established a foothold in the city can make integration easier. But as Lenard noted the consequences, though, might be problematic in terms of representation and equality, because their localities' interests will not be represented in the council. An egalitarian should object to practices of segregation which are clearly the result of policies or norms that have segregation as their aim. It would be a gross violation of the idea of a city of equals if a city had regulations that separate populations and influence a person's chance to reside in any part of the city based on their ethnicity, nationality, race, religion, and so on. But even without explicit regulations, as Anderson (2010) and Tilly (1998) argue, there are some practices of segregation, which are the result of entrenched norms, rather than policy or market transactions. This can happen in residential matters—where people want to move to areas that have more people 'like me' or get to hear about apartments available to rent through family and friendship groups. And patterns of non-residential segregation, or segregated social mixing, can also emerge through norms of behaviour. One obvious example is habits of avoidance, such as avoiding parts of town where out-group members gather, or not attending football games which the out-group members attend, and so forth. Municipal museums sometimes distinguish between what counts as high art and what does not, and therefore (unintentionally) between art of one group and art of the other group, and therefore between members of these groups (Kirchberg 2015).

So while we pay a lot of attention to spatial parameters for inequality in the city, we are careful about the limits of what can be deduced from them. For another example the Dutch sociologist Gwen van Eijk (2010) wished to examine the hypothesis that because information and ideas are spread in social networks and because these information and ideas can help to reduce inequality (as some of this information is helpful in that sense, such as when information about job opportunities is shared), then neighbourhoods as mixed networks will help reduce inequality, and if they are segregated then the opposite will happen. But her study did not sustain this hypothesis. She found that neighbourhoods are not as important as many assume in shaping and constructing social networks. Van Eijk is probably right. When we think about ourselves we can see that neighbourhoods are meaningful, yet their influence on our lives are limited. How many of your friends reside in your street or neighbourhood, and how many outside the neighbourhood? Do you work with people who live next to you? Does your family live in your neighbourhood? Presumably the answers to these questions are at least mixed. So neighbourhoods are not always that important in terms of networking. But moreover, neighbours are not necessarily the people to whom we compare ourselves when we think of equality. Despite the idiom 'Keeping up with the Joneses', which suggests that we do think of our neighbours as the benchmark for status and equality, people also tend to compare themselves to friends, to people who studied with them, to colleagues at work, to other family members, and therefore not necessarily to their neighbours.

At this point it is worth returning to and reconsidering the MTO (Moving to Opportunity) experiment discussed above in the context of Elizabeth Anderson's argument for the 'imperative of integration'. MTO was a randomized experiment conducted and sponsored by the Department of Housing and Urban Development of the US Federal Government in the 1990s. Our discussion is based on some studies of this experiment, and especially Chetty, Hendren, and Katz (2016). The idea was quite simple but very impressive. Four thousand six hundred low-income families from five cities (Los Angeles, Chicago, Baltimore, Boston, New York) who had been living in the same neighbourhoods were randomly assigned to three groups. Two groups were offered housing vouchers. The third group was the control, and remained where it was, not receiving any voucher. As for the groups that did receive vouchers, one group could use the voucher to rent anywhere in the city, whereas the second group could use the voucher only in a neighbourhood which was considered more affluent than the one they had been residing in.

Researchers studied the impact of this experiment on the housing, earning, and education of the families up to twenty years after the experiment took place. Obviously, if we take the simple, determinist, view that location and neighbourhood are the key parameters when it comes to progress in life, then there is reason to expect that all children who moved to more affluent neighbourhoods would do better. But the results are much more complicated and suggest that human capital as well as other social parameters can play an equally important part.

It turned out that moving to a more affluent neighbourhood when young (before age 13) was associated with positive impact, such as increased college attendance and earnings and reduced single parenthood rates, compared to the other two groups. But, alarmingly, it was also found that children who moved when they were already 13 years of age or older did worse than their friends who remained in the high poverty neighbourhood, perhaps because of disruption effects (Chetty, Hendren, and Katz 2016). At least three explanations of this difference are possible. One is that it is more difficult for older children to integrate into a new neighbourhood, and therefore the potential benefits of an 'improved' space will be lower. The second is that their personalities were already shaped to the spaces where they spent their early years, and they didn't 'fit' the new space, which implies that change of space had a negative effect. And the third is that they did not benefit from the move because they longed for their friends and comrades from the previous neighbourhood, in which case space was positive in creating a sense of community, but not necessarily in helping individuals who are transported from a place where they have a strong community to one where they have none. It is plausible, in fact, that all three mechanisms were in place.

To make this even more complicated, it seems that results also differed between gender groups. Girls in general benefitted more than boys (National Bureau of Economic Research n.d.). Another interesting finding which does point to the importance of location and space was that parents in families who moved to low-poverty areas had lower rates of obesity and depression (National Bureau of Economic Research n.d.). So what we see is that location matters and mixing matters but not to all, and to some it is counterproductive.

Some of these findings have to do with the intimate social ties that the adult poor had with each other in their original neighbourhood, which, when broken, had detrimental effects. This sustains what John Bird, who in earlier life had been homeless, and a rough sleeper, but nevertheless went on to found the magazine *The Big Issue* and now sits as a member of the UK House of Lords, told us in a previous study (Wolff and de-Shalit 2007). Bird described how he had lived in a run-down neighbourhood, where despite high rates of unemployment and poverty, people survived because of mutual help and a sense of community. However, when the authorities wanted to improve their accommodation, they offered the residents alternative places for the two to three years during which the place was renewed. Alas, when people left their immediate communities, they lost their social ties, and many returned as drug addicts, chronically unemployed, or even became homeless. When those who returned to live in the renewed neighbourhood came back, both they and their neighbourhoods were transformed and, they could not settle. So, according to Bird, while space had been positive in giving people a sense of place and community, it could not cure them once they had gone through the shock of displacement and returned in a perplexed state of mind.

So what weight should spatial dimensions be given when we think about the idea of a city of equals? Given our particular interest in relational equality, debates about Contact Theory, which is often associated with a seminal book by Gordon Allport (1954), is especially important for us. Contact Theory suggests that interpersonal contact between individuals from different groups can reduce prejudice. For example, living in segregated and desegregated housing units was compared, and it was claimed that there was a correlation between attitudes that white residents held and whether they were living in segregated housing units (in New Jersey) or desegregated housing units (in New York City). The latter were significantly less prejudiced. But Allport found out that in other cases things were more complex. Contact can yield more openness but also it can do the opposite. Analysing all these cases, he suggested that four conditions are necessary to reduce prejudice: members of the two groups enjoy an equal status; they all share common social goals; the members of the two or more groups do wish to, and do in practice, cooperate; and there is some institutional support for contact, at least in the minimal sense of there being no regulations prohibiting it, and, beyond that, there are policies which encourage it. Also, many years later it was established that contact had a positive effect on individuals even when they were not the persons inclined to choose contact over no contact (Pettigrew and Tropp 2006). How does the contact work? This is quite intuitive: by meeting the 'other' and learning about him or her, individuals develop empathy to, and lose anxiety about, the other.

However, notice that Contact Theory concentrates on individuals and their behaviour. In the last decade, urban political scientist Ryan Enos has argued that we need to consider those individuals' acts as conducted within the context of their groups. This makes sense as individuals experience their belonging to groups, such as women, African Americans, people from LGBTQ+ communities, in whatever they do; and second because spatially they are often separated to some extent, especially when we think of ethnic and racial groups in cities. Thus Enos (2014 and 2016), Enos and Celaya (2018), and Enos and Gidron (2016) argue that the picture needs to be deepened. Basing his theory on big data research as well as both designed and natural experiments, Enos argues that geographical space creates psychological space which creates political space. Because groups live separately, they perceive the 'other' in a certain manner. Space is used to help people map the social world in our minds. We tend to identify with those living around us, for example. Spatial separation ('here and there') therefore prima facie leads to social separation ('us and them'). We also ascribe to people who live next to us characteristics which are similar to ours, and, more importantly, to all who live in different areas, we ascribe, often subconsciously, what Enos calls similarity, that is, the feeling that all those I don't know are similar to each other in certain characteristics. However, a key claim in Enos's theory is that three parameters determine the level of trust and cooperation with members of other groups: the level of segregation, the size of the two groups, and the proximity—the physical nearness of the two groups and the likelihood that a member of one group will meet a member of another group. Size is a key issue here. It relates to the relative size of each group within the two groups. So, suppose that in a city there are 100,000 members of group A and 200,000 of group B; the size of B from A's perspective is 2 whereas the size of A from B's perspective is 0.5. Enos argues that everything else equal, in a situation whereby the two sizes are close to 1 the tension will rise, compared to a situation whereby one of the sizes is small. As an example, consider Enos's RCT experiment (Enos 2014).[5] For two weeks Latinos were asked to visit several train stations located in areas which demographically were rather homogeneously white. This created an impression as if all of a sudden there were many immigrants in the area. Other train stations in the area served as control. Passengers who were 'regulars' in these stations were asked about their political views about integrating immigrants and about Latinos and American identity. The results are clear: the surveyed persons who were exposed to the presence of many Latinos all of a sudden tended towards more conservative views about migration and were more open to excluding policies, compared to those in the stations were Latinos did not show up.

The rationale, according to Enos, is that when we see members of the other group in big numbers, we are more likely to perceive them as a threat and as different than when we see them in small numbers. Notice, that unlike contact theorists, Enos does not focus on the experience of interaction between individuals, but on interactions which affect (sub-consciously) our cognition and our perceptions of the other as members of groups. So contact between members of different groups in the context of segregation does not necessarily lead to more understanding, but, actually, might lead to the opposite, to more fear and even hate. In opposition to contact theory, according to Enos, close contact combined with spatial segregation along racial lines is likely to reduce city dwellers' readiness to embrace egalitarian policies and policies that invest in the well-being of the 'others'.

[5] We thank Itamar Alroey for drawing our attention to this important research. See also Alroey (2022) for a profound discussion of Enos's theory.

Whether Contact Theory or Enos is right, what is important in our context here is that segregation matters not only in economic terms, including employment and housing, but also in the way we perceive others. It therefore influences the most basic question: do we even care about inequality, and want to show good will and wish to limit gaps between members of different groups in the city? Both theories claim that spatial separation affects our political values. Which leads us to the discussion of theories of justice in the city.

2.5. The Just City: Towards a More Holistic Notion of a City of Equals

Several writers have explicitly taken on the task of deriving an account of the just city. We mentioned above that David Harvey in his book *Social Justice and the City* (Harvey 2009 [1973]) initially draws on Runciman and Rawls to offer an account of social justice in the city, appealing to need, contribution, and merit in a weak ordering (Harvey 2009 [1973], 100), although Harvey's attention to liberalism soon gives way to a more thoroughgoing socialist critique of the city in capitalist society. Harvey's liberal account is brief though complex, moving between distribution to individuals, through groups and territories, with some aspects of distribution directly to individuals and some mediated through territories. Harvey's notion of need is fairly conventional, although he is keen to emphasize that it is socially variable. Contribution as a key factor when an institution distributes access to its goods is the idea that those who work in ways that benefit more people have a higher claim to resources and services than those who benefit fewer. This seems not, however, to be a claim about intrinsic desert, but rather the beneficial effects of placing assets where they have the greatest multiplier effects, which, he notes, comes close to traditional concerns with economic growth, but in his schema is secondary to need. Finally, 'merit' is understood as a type of special sacrifice, so that, on an individual basis those, such as miners, who have to expend special effort to work should be rewarded. But in his application of this idea, it takes a territorial or environmental—by which he means spatial—twist: those who live in places that present special difficulties (such as in flood plains) should receive special treatment. How, then, does this differ from need? It appears that the main consequence of this criterion is negative: if you have chosen to live in a flood plain when other options are available, then your need-based claim is diminished in the case of a flood. Hence, for Harvey, merit takes the role of choice or responsibility in other theories.

As noted, Harvey's theory of social justice in the city is sketched rather than fully elaborated, and he notes the difficulties in moving from a sketch to concrete proposals. However, having got this far he neither elaborates nor attempts to apply his theory, so it remains somewhat under-developed. He did claim, though, in a more recent paper, jointly written with Potter (Harvey and Potter 2009) that in order to genuinely address issues of injustice in the city one needs to theorize outside the frameworks of liberal theory and capitalist society.

One of Harvey's examples of theorists who remains confined to theorizing about injustice in the city within the liberal theory is Susan Fainstein and her book *The Just City*. Indeed Fainstein's contribution to the question of justice is one of the most influential and cited works. Fainstein draws on three values: equity, democracy, and diversity, and much of her analysis brings out the potential tensions between these values, for if democracy amounts to majority rule then equity and diversity could be diminished. However, her interpretation of these values is as follows: equity implies that gaps in income are minimal; democracy implies that no person is excluded from the opportunity to influence the political process of self government; and diversity implies that no person is excluded because of her ethnicity or religion.

She applies her analysis to three cities, New York, London, and Amsterdam. In each case she uses a series of major development schemes as case studies to reveal the nature of the cities she discusses. She judges that Amsterdam does better on the three criteria (equity, democracy, and diversity) combined than New York or London, which is intuitively very plausible, of course. However, although her work is full of insightful empirical analysis, it is limited in a number of respects. First, she does not consider whether a special theory of justice is needed for considering justice in the city. Instead, she refers to works on justice in the state to analyse the city, even though these are different social and political institutions. Second, her own account is not developed in detail and hence her judgements remain largely intuitive (though not implausible). Third, her case-study methodology means that her focus is primarily on urban planning and the built environment, rather than encompassing factors such as the experience of day-to-day life (although in fairness these are mentioned in the discussion of cases). Of course she is free to pursue her analysis in the terms that seem most fruitful to her but we mention this because we hope to show that additional aspects of equality should be considered when concentrating on the question of the nature of a city of equals. Fainstein's goal of applying principles of justice to cases of urban planning, as an alternative to the neo-liberal model of planning, which subjects the process of planning to deregulation, privatization, and prices sets by markets, was highly important. Epting's distinction between 'urban planning' and 'city planning', is another way of capturing Fainstein's approach. By city planning Epting means an approach to cities that derives from understanding what it means to love the city, whereby 'all urban dwellers can see their power in the thinking behind the process of changing the city (. . .) through discussions, debates, compromises, and civil arguments' (Epting 2023, 78). Fainstein wanted to shout loud and clear: there should be other considerations, and we should not leave everything to the market. She insisted that we pay attention to how space is a product of social processes, how interests shape planning, how property markets, the built environment, are not neutral terms, how the idea that we can separate the aesthetic from the political is false, and that therefore planners, architects, and designers should be concerned with social issues. As an example, the reason Amsterdam's social housing was spread all over the city and not concentrated in separated neighbourhoods was social; it derived from the values of equity and diversity, and was therefore just.

Yet Fainstein limits the scope of her study to the context of planning (broadly understood), whereas our argument is that while planning is an important aspect of equality and inequality in cities, it is only part of the picture. It might preserve, even create, inequalities, and though it can also reflect the city's egalitarian values, it has limited influence in bringing about these values. True, it can help the city integrate, equalize, and compensate for disadvantages; but it is only one part of a portfolio of measures available to a city. Fainstein is right that planning policies should take into account the interests of employees, not merely owners and employers, and that megaprojects should be subject to scrutiny so that they provide benefit to low-income people in the form of employment provisions as well as in other ways. These and other suggestions of hers are highly welcome; yet they do not exhaust what city dwellers intuitively think of when they think of an egalitarian city (as we shall demonstrate in Chapters 3 and 4, where we analyse the interviews we conducted) and as was suggested more recently by Talja Blokland (2017; 2023), who perceives justice in the city in terms of a strong sense of community, not necessarily regular face-to-face interaction but rather non-durable, fluid encounters in urban public spaces (see also Valentine 2008). Thus, Blokland argues that in the city, we encounter people whom we do not know and whom we might not know or meet again in the future. Yet, through this encounter, we learn a lot about opportunities in the city, and about ourselves in the urban environment. Moreover, this encounter can enable us to help somebody in need, learn about those marginalized, and 'engage in moments of sociability'.

Detailed explicit philosophical reflection can also be found in the concluding chapter of Iris Marion Young's *Justice and the Politics of Difference* (Young 1990) which, incidentally, begins with a citation from Jane Jacobs. The nature of Young's project is not made fully explicit although she does say that she wants to develop a 'normative ideal' of city life. Hence, we will take her as proposing an account of a 'good city' which is not necessarily the same as a city that embodies the egalitarian spirit. Nevertheless, having in mind her body of work, for Young, naturally the two things will come close. For Young, the city, or we should say the ideal city, is defined sociologically: individuals and groups interact in places, spaces, and institutions to which they all feel they belong. What distinguishes the city from small towns and villages is that these interactions do not work against the uniqueness of each individual and/or group. In other words, the tension that drives her project is the combination of individual and community in the city, for she is keen to avoid any form of communitarian merging of identities, especially on an involuntary basis. Young emphasizes the idea of 'public vitality': the city, for Young, should be exciting and take one out of the routine, and in that sense, she says, is 'the obverse of community' (Young 1990, 241), which she understands in terms of a type of semi-stultifying conformity. She notes, 'As a normative ideal city life instantiates social relations of difference without exclusion' (Young 1990, 227; see also Wolff 2017). Face-to-face relations are always mediated, and a city is made up of relationships between strangers, though Young stresses that alienation and mediation are not the same. It seems that alienation is avoided not always on the individual level—that is to say that individuals may still be strangers to each other—but mitigated because they all feel that they belong to this big social project, namely their city. The many social networks, community groups, and other aspects of civil society in the city sustain a sense of belonging and can reduce feelings of alienation. It also helps to create heterogeneity and inclusion.

Young builds up a picture with four main themes, to which she refers as the city's virtues:

1. Social differentiation without exclusion which is created by many and varied social gatherings and interactions, including individuals moving from one gathering to another.
2. Variety of institutions, restaurants, cafes, places of meetings, and a variety of uses of the city, which create a sense of place and a safe space.
3. Eroticism by which she means, not, we think, anything directly involving sexual contact or approach, but the pleasure of being drawn out of the routine into the unexpected and thrilling.

4. Publicity, which she understands as the availability of public spaces and of many interactions which include disagreements, debates, demonstrations, and forms of voluntary activity, all of which are fertile for the city.

Young notes that contemporary cities are very far from instantiating this ideal, but precisely because of this we find Young's picture very inspiring and appealing, and in some respects it overlaps with the analysis we will present. However, our own question is much more focused on the idea of embodying the egalitarian spirit, rather than a 'normative ideal' of the city; and in drawing on the urban studies literature and our own interviews, we will present things in a rather different way. One reason we find Young's approach so insightful is her understanding of the city as space of relationships, where the quality of relationships constitutes an 'ideal' city or, as we would put it, a city of equals.

Perhaps as a mirror image of Young's ideal city, de Silva et al. (2021) studied the city of Manaus in Brazil, pointing to how important relationships are for making a city egalitarian, in the most basic sense of the term, namely that people can survive without fear, that they have self-esteem and that they feel part of the city, and emphasizing the potentially devasting effects if such relationships fail. Making use of the theory of corrosive disadvantage[6] that we set out in *Disadvantage* (Wolff and de-Shalit 2007) as well as Bourdieu's theory of symbolic violence, de Silva et al. argue that the very high current levels of physical, as well as symbolic, violence in this city particularly affects the least advantaged who reside in extremely violent neighbourhoods and have to sacrifice some functionings (affiliation, ability to work) in order to protect the critically important functioning of bodily integrity. The authors cite female interviewees who gave up their jobs because of the high risk of robbery as soon as they received their wages, or being attacked when they leave the workplace on their way home. Others described how they gave up socializing because being outside increases vulnerability to violent crime. Interviewees said they chose social isolation to minimize exposure to risks of violence. A further effect is that few informal social institutions develop in the neighbourhood, and those that do are very weak (de Silva et al. 2021, 7). The authors argue that 'distrust of others and reducing social interactions erodes the capacity for affiliation, which is based on trust, belonging, respect, and equity' (2021, 11).

[6] We defined corrosive disadvantage as a case whereby disadvantage in one functioning leads to disadvantage in others.

Another very interesting step forward in tying urban justice and equality in the city to relational equality is provided by van Leeuwen (2020), which we mentioned at the opening of this chapter. He claims that a recognition-theoretical approach such as a modified version of Honneth's theory of recognition, paying more attention to ways in which people differ from each other, should be relevant to cities as it is already both spatial and relational, and includes diversity. These spatial and relational elements combine with diversity to comprise what van Leeuwen regards to be the three aspects of urban justice which distinguish it from justice within the state, based on issues of basic rights and wealth distribution. On the city level, he claims, questions of justice concern mainly the way urban space is organized, and what it expresses. The just city, van Leeuwen claims, is a city where space is structured to meet the demands for recognition. This is an important step forward and ties in with our account of a city of equals, and has influenced the direction we take in the next chapters. We also aim at understanding what such recognition, or what we define as having a secure sense of place, consists of.

Finally, not everybody holds that inequality in the city is, morally speaking, prima facie bad. First, we must remember that often inequality and poverty in cities is not always the result of any policy by the city, but rather the result of the fact that the particular city is attractive to poor and disadvantaged people. Glaeser (2012) argues that cities are therefore not to be blamed for the resulting inequality and poverty. Well, to this we answer that while it is true that cities attract the poor, we are not in the business of blaming. Instead we hope to offer materials that will help city officials develop ways of minimizing inequality. We are, to repeat, especially interested in how a city responds to inequality and poverty, whatever their causes. For this we claim, we need first to understand the complex nature of inequality in the city. Although we will set out an account of a city of equals, this is not intended as some sort of blueprint or ideal notion of equality in the city. Instead, we identify clear cases of inequality in the city and learn and build from these cases and issues in order to draw out the main determinants of equality and inequality in the city. We want to understand what it will take to move cities in the direction of being cities of equals.

Second, although our topic is inequality in cities, nevertheless it has to be looked at within the context of the state, in that it would be problematic if all the cities within a state are internally broadly equal, but externally are highly unequal, with some consisting only of rich people and others of only poor people. Rae (2011) we noted above, argues that inequality within the city, if it is not too radical, implies that the population is not overly homogeneous and that rich and poor live in the city together, rather than in separate cities, in segregation. This, Rae continues, affects not only inequality and equality between cities, but also the well-being of the least advantaged within the city. Margaret Kohn (2011) also argues that human plurality is good for the city, though she argues that this is because of the democratic and social utility that emerges from the meeting of different populations. We cannot deny that there is some truth in these arguments; but while the interest of the poor in living with the rich is obvious, at least in the best cases, it is less apparent what the interest of the rich is in living with the poor. This becomes clear in the light of Macedo's argument (2011) that the rich prefer affluent cities, those that are less mixed.[7] So we accept that some rich people would prefer to leave a city if the city does not permit them to enjoy their wealth and privilege. Perhaps we should concede that human nature is what it is. We are not expressing any view about whether it is ever possible to transcend inequality, and in that sense we are realists like Rae. However Rae also claims that it is morally good to have some inequality within the city because otherwise inequality will fall between the city and the suburban penumbra. (Rae 2011, 105) Rae appears to assume that some form of inequality is inevitable at state level, and a society's main policy choice is simply where that inequality should be located. We do not share these assumptions, but in any case our topic is primarily what it means for a city to be a city of equals and here take no position on whether inequality at state level is inevitable.

2.6. Conclusion: The Many Dimensions of a City of Equals

Many theorists of all disciplines have been greatly concerned about the injustices they perceive in the city. The most visible manifestation is spatial, which very often, though not exclusively, is correlated with race and frequently economic class, and the intersections between race and material income and wealth have troubled many, especially as they also correlate with inequities in housing, education, jobs, and transport, as well as racial and gender discrimination. All of these factors are critically important and will figure in our own analysis.

However, to summarize, we find two significant areas where more work is needed. First, discussions have typically focused on income and on space, and, to a degree, on processes of decision-making, but much less on what it feels like to live in a city in relation to others. Second, most, though by no

[7] We would like to thank Tal Banin for this point.

means all, work on justice in the city has not considered the city to require particular treatment as a special topic but has applied a theory of justice initially designed for conceptualizing justice at the level of the state. So in the following chapters we will attempt to address these gaps, first by setting out some of the findings of the interviews we conducted and then in presenting our account of a city of equals.

3

Interview Themes and Results, Part 1

3.1. Our Starting Point

As discussed in Chapter 1, we did not approach the interviews with a blank piece of paper. Instead, a considerable period of reflection and reading preceded the interview studies leading to the formulation of a basic approach to the topic, which we sought to test, challenge, modify, enrich, and ultimately specify, on the basis of the interviews. The core idea, which seemed to us the essential basis to develop in much more detail, is that a city of equals is not so much a city in which there is equality of income or wealth, or homogeneity of population, but rather one in which it is possible to make a meaningful urban life whatever your economic resources, and your personal characteristics and identity such as gender, race, religion, ethnicity, disability status, age, sexuality, values, interests, preferences, and so on. In a city of equals you are made to feel that you matter, and, most essentially, that you belong. Summing up a thought many interviewees expressed in their own way, we boil this key idea down to the sense that you are proud of the city, and the city shows that it is proud of people like yourself. And you also have the sense that a meaningful life, mattering and belonging are available to all, and not only to a select elite, or even the majority. Instead, there should be a welcoming place—a sense of belonging—for everyone.

As an example, which is anecdotal, and at the same time telling, consider this. Many residents of Amsterdam are aware of their city's nickname, *Mokum*, and there are several galleries, restaurants, and cafes that carry that name. But it may be that fewer are aware of the reason for this. In 1796 the Jews in Amsterdam were emancipated; there was a new regulation, an enactment of equality for Jews, equivalent to what we would now term city naturalization. This made a strong impression on the Jews in the city. For example, they decided to introduce the Dutch language into their schools. But more interestingly, throughout the years Amsterdam's Jews started to refer to their city as *Mokum*, which in Yiddish, the language spoken by many European Jews at that time, means 'place' and also sometimes safe haven. Indeed, their feeling was of now having a place. Until then Jews in

City of Equals. Jonathan Wolff and Avner de-Shalit, Oxford University Press. © Jonathan Wolff and Avner de-Shalit (2023).
DOI: 10.1093/oso/9780198894735.003.0003

Europe thought of themselves as living out-of-place, in the diaspora. They were not welcomed, they were not allowed to own and cultivate land, and often suffered from violent attacks and pogroms. They therefore prayed to return to Jerusalem (Zion). Amsterdam's decision to legally acknowledge their equal status, which followed many years in which in practice they were tolerated and even welcomed, made the local Jews feel at home and they therefore called the city 'place', indicating a sense of settledness, perhaps even integration, belonging, not being out of place (Ostow 2005; de-Shalit 2021).

3.2. The Results of the Interviews

We undertook the interviews over a period of four years, and we discussed and reflected on our central claim and how to develop it into further themes as we considered the transcripts and summaries of the interviews. Therefore, there was a development both in our ideas and the nature of the interviews as time passed, allowing us to refine our understanding and use of the interviews, rather than sticking to the same script in every case. In what follows in this and Chapter 4 we will weave our interview results with observations from the literature, sometimes going beyond the literature review of Chapter 2, to compare the claims of theorists with the experiences of our interviewees.

There are a number of ways in which we could present the interview materials. There is far too much to print transcripts, or even summaries, of all the interviews. Accordingly, in this chapter and the next we will present and discuss extracts that we found most illuminating. We shall discuss the themes as relating to the city's four core values which we mentioned earlier: non-market access to goods and services; sense of meaning; diversity and social mixing; and non-deferential inclusion. Of course, some themes fall under two or more core values. Subsidized public transport is a good example, as it is clearly about non-market access to services but also about enabling people to visit friends, attend schools and universities, commute to work, and so on, all of which relate to meaningful life in the city as well as non-deferential inclusion. Thus, the allocation to some themes to particular values could be arbitrary and changeable.

Also the break between this and the next chapter is somewhat artificial, in that we have too much material to restrict ourselves to one chapter, but we could have presented the material in various different orders. Hence this and the next chapter should be treated together. However, in this chapter we primarily look at the first two core values: non-market access to goods and services, and meaningful life, which can be characterized as dealing with the well-being of the individual as an urban self: it is about what is so attractive to many people about the city and urban life, and how a city of equals should enable each and every of its resident to flourish and enjoy urban life. The next chapter, Chapter 4, concentrates more on how the city as a social unit respects the differences between city dwellers, hence diversity and social mixing, and non deferential inclusion. We will then, in Chapter 5, draw on the interviews, as well as the literature review and our own reflections, to develop our conception of a city of equals as providing a secure sense of place for all, relating it to the four more specific, but also more comprehensive parameters, which we will present as core values of a city of equals.

3.3. Relational Equality

Before we move to the thematic breakdown, however, it is worth framing the discussion in terms of how our interviewees approached issues of equality. Of course, many were concerned with material distribution, especially as it affects such things as housing, local amenities, and access to, and variety and frequency of, transportation, to which we will return shortly. But material factors far from exhausted concerns about equality. We have several times suggested that not enough has been written about relational equality, or social equality in the city, and in our interpretation of our interviews, nearly all city dwellers think about equality in the city in relational terms. Moti, a 63-year-old resident of Tel Aviv, says:

> They say Tel Aviv is the culture city, but I can't find anything cultural in it. It lacks compassion and benevolence. Social alienation here is the worst. If a neighbour dies, no one will know until he'll start stinking. It's not how it used to be, when you could knock on your neighbour's door. (. . .) In my work I've seen people in the worst situations; and yet there is no volunteer work here. (. . .) I don't understand where this kind of behaviour comes from, where are their hearts?

Jenny, aged 28, also in Tel Aviv, thinks that equality is not only about what we offer to the more disadvantaged but also about the concern and respect with which the city provides its services:

> I can really tell the difference between the way you get access to services in different areas. When you go to the welfare bureau in south Tel-Aviv [the less affluent neighbourhoods], you have to wait in line without even having a glass of water. There are no coolers there, and obviously you can't make yourself a cup of coffee.

And then she adds, in terms that we would say picks up on the critical importance of relational equality:

> We need to be sensitive to the dignity and self-esteem of those who need help. (. . .) I think that is exactly the problem. I'll give you an example. There is this soup kitchen down south that is decorated to look just like a restaurant. And you tip the waiter there, even just with one Shekel [about a third of a US dollar], so you wouldn't feel that you're getting free food. I think that the fact that one is eligible for welfare payments (. . .) he still has the right to feel equal. He deserves a glass of water. That is why when we [at work] decorated our youth centre we bought new furniture, instead of just getting some donations, because we wanted the guys to feel like they deserve better.

Yet our interviewees were, in some cases, deeply concerned by economic issues too. Valentina, 37, from Rio de Janeiro says:

> I'm horrified by the economic inequalities that I see here. There is a great distance between rich and poor (. . .) extremely wealthy neighbourhoods with everything, such as Ipanema, and extremely poor neighbourhoods without anything, such as the favelas.

However, she immediately adds that one key manifestation of this inequality is that poor people cannot reach the two most important areas: the city's centre and the beach because they are very badly served by public transport. Indeed, in many cases, even when they referred to material inequality, interviewees immediately framed them in relational terms. Thus, Valentina calls Rio a 'city of division': 'I have a close black friend. He said to me that when he needs to take a taxi in town it is very rare for a taxi driver to stop and pick him up. This is so wrong.' (This echoes Arthur's (35) description of the same city, Rio, for similar reasons, as 'a city of brutality'.) She then goes on to point to what is now commonly termed intersectionality in the city:

> The coloured people are not only a minority, ethnically speaking, but are also poorer, and work in the less sought-after jobs. Another example is when you go to the shopping centre, you'll see white people buying, black people working, serving them. (. . .) There are different kinds of shops: some for the rich, others for the poor.

The highly visible economic divide was picked up by other interviewees in other countries. Alex, a 41-year-old male interviewee in London says:

> I've just been into the Isle of Dogs, you know, Canary Wharf, and I went to the supermarket there. It's just amazing, you find very poor people there in the south

> of Canary Wharf. And it's just interesting to see the poor and obese people in the supermarket, and then, you know, just around the corner the fancy people are getting out of their Bentley.

Alex, identifying himself as a lower-paid worker, clearly felt that his life opportunities were restricted by lack of money, and the high cost of living in London, but nevertheless, he was able to find some coping strategies:

> £30k really isn't enough for anything. I couldn't buy a house, couldn't have a family. And who are you gonna seduce as a 41-year-old with a salary of £30k a year, you know? So I'm doing a lot of things that help me cope with the stress of being in London, like 10 yoga classes a week (*laughing*).

There is a lot of truth in what he says. The fact that yoga classes are open and accessible, often cheaply, to many people whose salary is low, and the fact that in these classes or other such amusements they meet people from different backgrounds makes one feel that inequality in the city is not as harsh as it could be. In his book *What Money Can't Buy* (Sandel 2012) Michael Sandel fondly recalls the times when, he claims, people of all classes met each other in the baseball stadium—they shared the queue to the ticket office, they paid more or less the same price, lined up for the toilets together, and they bought the same lousy hotdog after the game. Nowadays, he regrets, they are separated in different sections of the stadium, as a market rationale has penetrated to sports events. So the gap between them is not only bigger, it is also seen and felt every minute, as more and more goods and services are subject to the power of money, which, in turn, diminishes the sense of community.

But there are many nuances to the picture in the city. Alex notes:

> Well, the thing is poor people in London, they may still have an advantage of living centrally, you know, it's the one advantage. You're very poor, but you're living in Zone One (if, of course, they can afford it . . .)

Alex's point is that in London highly subsidized social housing can be found all over the city, even in some of the most lively and affluent parts. Our aim is to capture at least the essential elements of this complexity.

Valentina and Alex, cited above, do mention issues of income and wealth. However, they are far from representative of our interviewees, very few of whom mentioned income gaps as a matter that concerned them in particular when they think about well-being in the city, and how access to it is distributed between city-zens. This is also consistent with Martina Löw's argument (Löw 2013) which has become a type of touchstone for us in this project, that city-zens regard their cities as 'entities of meaning'. Löw's argument can explain why the type of attachment people have to their cities differs from that they have to their country or state and why their expectations from the city differ from their expectations from the state. The interviews confirmed Löw's perception of cities, as people time and again described their cities in intimate terms. As the philosopher Avishai Margalit, who was born, raised, and lived in Jerusalem once said, 'I don't like Jerusalem, but I love it' (Bell and de-Shalit 2011, 52).

3.4. Themes that Relate to Non-market Accessibility to Goods and Services

We begin our detailed review of the themes that were expressed by the interviewees referring to those themes which express the core value of non-market access to goods and services.

3.4.1. Spatial Dimensions of Integration, Segregation, and Their Consequences

Integration and segregation could naturally be discussed with reference to the core value of diversity and mixing. But our point here is that while this is true, it is also interesting and important to see how segregation might affect city dwellers' opportunities to access the goods and the services that the city offers. We therefore wish to begin with a few words about how the interviewees described integration and segregation as related to diversity and social mixing, but continue with how this affects accessibility to goods and services, and how this might, but should not, be a function of the market and one's wealth.

Within the academic literature there is a distinction commonly made between two types of group segregation: clustering and isolation. Clustering concerns the concentration of distinct socio-economic groups across neighbourhoods. Concentration, and therefore clustering, is high if all the members of a particular socio-economic group live in the same neighbourhood, lower if they live in several neighbourhoods, and lower still if they are not associated with any particular neighbourhood. Isolation is related to how unlikely it is for a member of a group to meet a member of another group. Although related, the two concepts are different. If, for example, a socio-economic group is spread between several locations, but highly concentrated and self-contained for work, leisure, religious worship, and education in each, isolation can be high, but concentration lower (see OECD 2018).

Both clustering and isolation were mentioned, albeit mostly in other terminology, by our interviewees as significant contributors to inequality in the city. For example, several interviewees in the southern neighbourhoods of Tel Aviv described the clustering there of minority groups, poor people, and undocumented immigrants who work in Tel Aviv, whereas in the north, 'I want to see what will happen if an illegal immigrant asks to rent there—the police will immediately be invited.' Isolation—the unlikeliness of meeting a member of another group in the city—appeared most frequently in our interviews not so much as consequence of pure material factors of planning and the location of housing and amenities but as intertwined with more relational factors. For example, when the dominant culture is of suspicion and avoidance of 'the other', then a member of a minoritized group will feel unwelcome in the parts of the city where the majority resides, and will tend not to visit there; in that case this person feels isolated. So, for example, interviewees we interviewed in Blackbird Leys, one of England's most deprived neighbourhoods located in the south of Oxford[1] told us that they simply avoid going to the city centre because people in Oxford are snobbish, they say, so they would feel unwelcome there.

Nevertheless, some interviewees see no injustice in the way the city's population is arranged spatially, even when there is some separation of more advantaged and disadvantaged populations. For example, R migrated from southern France several years before we interviewed him in Manhattan, New York City (NYC), in 2016. He was in his thirties and lived in Queens, NYC. He ran a tiny cafe (four people could sit inside) in the lower part of Manhattan. The espresso was really good. When we asked him what equality in urban life meant for him he said:

> What do we care about and what do we expect our city to offer? Safety for our kids when they walk in the street; having good friends around; good schools; accessibility to all kinds of places and attractions.

And then he added about his neighbourhood in Queens: 'People don't see me as different because I speak with a French accent, as so many of them are immigrants themselves.' When asked if there are other parameters which

[1] A focus group conducted for the purpose of previous research.

imply that Queens is such a good place he said it was close to the seaside, which was great for his family. Then he added something important: 'Basically,' he said, '*I don't worry* when I am in Queens: my kids are safe, I know that they have good friends, and I have good friends too. I don't care about access to the jazz clubs and the theatre; I care about access to shops, schools, and the seaside.' R. describes advantages and disadvantages in spatial terms, including how access to places as well as to his friends enables him to feel secure and not worry.

We return below to integration, inclusion, and exclusion, but now we wish to consider their material dimensions, including those that are often ignored. For example, public toilets, not so much discussed by philosophers, but vitally important for the experience of the city, are worth thinking about. They are typically built by the municipality and they are crucial for enabling people to walk outside their homes, to stay outside home or, for some types of employment, work for longer, to go shopping and just to stroll in the public space, especially for parents with young children, and, sometimes, older city-zens, but of course, ultimately, for everyone. A city without public toilets is much more problematic for those without the money to go to a pub or restaurant, or the confidence to walk into a hotel lobby, if they need to, in order to have access to private toilet facilities. Perhaps the worst affected are those homeless people who sleep on the streets, and are often not allowed to use lavatories in commercial places like restaurants and cafes. Thus, Jeremy Waldron suggests that 'If urinating is prohibited in public places (and if there are no public lavatories) then the homeless are simply unfree to urinate' (Waldron 1991, 315). Therefore,

> the generous provision of public lavatories would make an immense difference in this regard—and it would be a difference to freedom and dignity, not just a matter of welfare.
>
> **(Waldron 1991, 321)**

Mothers with children often learn which department stores or shopping malls have freely usable toilets. But where there is thoughtful public provision, rather than haphazard private provision, the city becomes more accessible to all. This is clearly and simply a spatial aspect of equal treatment, which was mentioned mainly by female interviewees.

A different type of material example was brought to our attention by an employee of eBay in a European city. The company, he said, has data about neighbourhoods that lack a post office where boxes can be posted or collected, or in which residents have to walk or travel some distance to reach a collecting point. The company, we were told, has less interest in advertising and improving its services in such neighbourhoods, as people are unlikely to use them because posting or collecting the boxes is a burden. The result is that residents of the neighbourhoods that are poorer and less well served by the post are furthermore discriminated against when they cannot use the service, which might have helped them purchase better value goods. And what is true for eBay presumably generalizes to other companies delivering goods to homes. Furthermore, those who live in large blocks of flats, without a concierge, are presumably more liable to their parcels going missing once delivered to the building. Hence they may be put off from purchasing online and again miss out on cheaper deals. And companies may be reluctant to send their goods to locations where they often get complaints of missing deliveries. Of course, these are exactly the circumstances in which many poorer people live.

Spatial inequalities were often noted in our interviews. As Sabine, an interviewee in Berlin said: 'Those from the East [Berlin] are strongly disadvantaged . . . they need to secure additional earnings somehow, on top of their pensions.' And as we noted in Chapter 2, the fact that inequality in cities has a spatial dimension has been widely observed in classic works in urban sociology and urban political science, such as Wilson (2012 [1987]) or David Harvey (2009 [1973]) who, for example, refers to 'the philosophy of social space', when he analyses the speed of change and the rate of adjustment in an urban system, or the redistributive effects of the changing location of jobs and housing. It has been also analysed in more recent empirical works, such as Mahadevia and Sarkar (2012) and Anderson (2010). These works, though, show how inequality coincides with exclusion or spatial segregation, whether imposed, in former times, by legal regulations, or by norms, or is the result of the markets, as we discussed in Chapter 2. We also noted that some critics argue that spatial differentiation does not always take the form of segregation in a negative sense, if the minoritized group, whether identified on a cultural or ethnic basis, wishes to stick to each other and live with 'their own' (Sundstrom 2013; Shelby 2014). We can see how, for example, ultra-Orthodox Jews live side by side with each other and how interviewees in Jerusalem told us that having a synagogue where prayers are conducted in the fashion they are used to is crucial for them when they choose where to live. For example, in a study parallel to our own, Jonathan (19) was reported as saying:

> What I mostly love in our neighbourhood is my synagogue; this is where I study, pray, this is my main place in life. I therefore try not to leave the neighbourhood.
>
> **(Ben-Dahan 2017)**

And yet, as Sundstrom argues, even if the segregation is initially voluntary, it will very often lead to growing inequalities between neighbourhoods or even streets, such as in the services they receive. For example, Shatilla, aged 50, from an Arab neighbourhood in Jerusalem, complains that residents there are discriminated against in services:

> There is no proper place for my children to play at. Nothing like the place we are sitting in right now [in the Jewish neighbourhood where they, she, and other Arab city-zens work]. I pay just like everybody else, but I don't get any of the services. They hardly take out the garbage in my neighbourhood, and the roads are terrible.

Differentiation in waste collection in Jerusalem has been confirmed by research by Issar (n.d.) and Bimkom (2012), an Israeli NGO of architects that monitors discrimination and harm to Palestinians' rights during the planning process. Of course, less regular collection of garbage or street cleaning will leave poorly served areas more heavily littered for longer, feeding cultural stereotypes about the tendency of poorer people to care less about the condition of their streets. Another form of spatial inequality concerns street planting. The planting of trees in cities, especially those of hot climates, to give shade from the heavy sunshine, has also been noted to follow patterns of wealth and deprivation, even though they are paid for out of municipal funds (Davis 1997).

Spatial factors have further significant effects on how people are able to experience their lives. First, spatial arrangements and city planning are a significant factor in city dwellers' capabilities, by which we mean their *genuine* opportunities (Wolff and de-Shalit 2007) to practice their functionings. By genuine opportunity we mean both the bare presence of opportunities, and also the real possibility of taking up such opportunities without sacrificing or risking other things of great importance. For example, if someone has to travel a long distance across town to go to dance class, we would question whether they have a genuine opportunity to take dance classes (even though it is a far better situation than one in which there are no dance classes or they are prohibited). Practicing one's talent in opera singing or karate is so much easier if there are opera clubs and karate studios near to home, at reasonable prices, and the same thing is true of securing health and the proximity of a free or heavily subsidized clinic nearby. This may sound trivial; but it is central to life especially for those living in the city on low incomes. Everyone's capabilities are conditioned by such spatial arrangements in their city, and people living in wealthy parts of the city are typically very well served. For poorer people it can be especially difficult to reach a dentist or a health clinic, as well as a library or a karate studio if the cost (in terms of money and time) of reaching the place is relatively high or even if the cost is not very high for the average person, but is a high percentage of a low income. In other words, the question is not only whether urban goods and services are available in the city, but the distribution of accessibility, or who can and who cannot use these services easily, and whose freedoms in the city are expanded and whose are limited. Consider how, when in June 2022 the American Supreme Court ruled against the federal constitutional right for abortion, there was much talk about how it would disproportionately affect poorer women, as many women would now have to travel from states in which abortion would become illegal to neighbouring states where it would remain legal, and the cost of travel is more of a burden for poorer, rather than more affluent, women.[2]

Many interviewees claimed that inequality in the city consists of dimensions which have some spatial character, and greatly affect accessibility in practical terms, *but go beyond segregation.* First and foremost, and especially in Rio de Janeiro, but also in other cities such as Hamburg, interviewees remarked on the spatial dimensions of exposure to violence. As we noted in Chapter 2 in relation to the Brazilian city of Manaus (de Silva, Fraser, and Parr 2021), when people are exposed to violence they typically avoid going outside, even to the point of not seeking jobs, so as not to be exposed to risks to their life and safety. Thus those among the least advantaged, who reside in places where there is a high level of violence, have to decide whether to sacrifice their functioning of working in order to secure their life and bodily integrity. Accordingly, even if facilities are available and close by, there is a serious lack of opportunity and capability if people do not use them because of fear of violence.

In many countries, police services are controlled by the city rather than the state, and the police is run by the local mayor. Accordingly a city of equals could make a difference to policing, by providing more of such things as greater police presence, and faster arrival of a police van following an emergency call, although how welcome such intensifying of services will be will also depend on how trusted the police are, which we cannot take for granted,

[2] See also Frediani (2015) and Frediani and Hansen (2015) for how 'the availability of infrastructure or characteristics of the built environment might compromise or facilitate individuals' abilities to enhance their well-being' (66). One of their examples is illuminating. The use of bicycles can help poorer people tremendously: to reach locations; find a job and easily reach their workplace; visit relatives, and even for leisure. But for this to happen the city needs to regulate and must also provide special arrangements for secure and easy riding. Indeed, we wish to add the example of Copenhagen, where a system of 'intelligent traffic lights' prioritizes buses *and bicycles* so that the time of reaching places by bicycles is reduced, and is even sometimes faster than the car (Davies 2016). We would also add the importance of facilities for safe and secure storage of bicycles especially for people living in cramped premises or high crime neighbourhoods.

especially in minoritized communities. But putting issues of trust to one side, even though often crime rates tend to be higher in poor neighbourhoods, in many cities police services are offered more intensively in neighbourhoods where the more affluent and politically powerful populations live. In Tel Aviv two interviewees complained that the police are often reluctant and slow to show up when they are called by residents of the southern, less affluent neighbourhoods. In contrast, they respond very quickly to any call by residents of the northern, more affluent neighbourhoods, for example, when an illegal immigrant is seen there and is considered a nuisance. And Valentina, an interviewee from Rio, reported that police officers will tend to be more violent towards poorer people.

These claims are consistent with many research findings. Often where poor minoritized communities reside there is under-policing and police neglect of their neighbourhoods (Ben-Porat 2008). Moreover, Joel Suss and Thiago Oliveira (2022) show (based on a study of data from London in 2019) that police officers more frequently stop and search members of the public in neighbourhoods where well-off and economically precarious people co-exist, and of course it is those who present as more precarious who are stopped. They claim that economic inequality is positively associated with Stop and Search incidence in those neighbourhoods. That the effect of neighbourhood on police activity can be even greater than race is confirmed also in research in the United States (Terril and Reisig 2003). So the provision of policing should not be influenced by the wealth and power of those receiving the service. We return to the theme of security later, when we discuss the value of a sense of meaning.

Policing is just one example, though, of the way facilities are not evenly spread throughout the city. Interviewees mention the frequency and variety of transportation which is available to different neighbourhoods (more on this later), or the number of cafes and health clinics or even supermarkets in their neighbourhood. There is evidence that people in poorer neighbourhoods consume more junk food because going to the supermarket is time-consuming as chains tend not to locate their shops in the poor neighbourhoods (Wolff and de-Shalit 2007). Ease of transport is central to many people's quality of life. For example, Maria (25), a Jerusalemite, says:

> I am happy with the transportation here. I can easily walk from my home to the bus station, and there are tons of buses there. I take traffic into account, but even so it doesn't take me more than twenty minutes to get to work. Without traffic it takes about six minutes. I know everybody likes to hate the light train, but I am fine with it.

The nature of public transport has far-reaching effects. Arthur (35) from Rio de Janeiro, we noted earlier, calls the city a 'city of brutality'; and he associates it with the fact that people from different background and classes do not mix, because not only do they live in separate parts of the town, but also the public transportation system is:

> terrible, (. . .) [the city's] north and the south are not well connected. [Hence] our population is so segregated. I mean that our public transport makes the interaction between those from different backgrounds and neighbourhoods even harder. . .Geographically, the city of Rio looks like a maze of hills. The city is winding. Lots of curves surround you, they trap you. It's hard to find a way out. Most white and upper-middle-class people live in the south of the city or even in the rich suburbs, while lower-middle-class and poor people are in the north or west of the city, in the favelas or very far away from the centre. Public transport for those living in the poor areas is really terrible, they're left far out from the centre and the beaches. Look at the metro lines! [Mostly concentrated in the wealthy south.] This makes our lives very difficult. I think twice when I need to visit a friend who lives in another neighbourhood.

This testimony is echoed in the inequality map of Rio de Janeiro (Casa Fluminense 2013), which uses twenty-three indicators and seven key themes to analyse inequality according to neighbourhoods in the city. It can be clearly seen that there are spatial differences, and, for example, that in the same neighbourhoods (in the north-west) we can see high percentages of people who spend more than an hour travelling each way to work, high homicide rates, low percentages of children at the age of 4–5 in pre-school and very low levels of people served by a sewage system.

Our interviewees were keen to discuss what is available to them and under what conditions. The first, and, for many, most pressing, issue is accessibility of basic services regarding nutrition and health. Gerd, 72 years old, was born and raised in Hamburg where he still lives, and he loves the city. When he is asked to say what is the best thing about the city he says:

> Everything is easily reachable and within short distances, doctors, etc. I always use public transport. I got rid of my car when I stopped working and I haven't regretted it for a single moment.

Torge (male, 25) from Hamburg also acknowledges that it is easy to reach a grocery store or supermarket everywhere, as well as cafes, snack bars, and bars. He laughs: 'in terms of gastronomy we are being looked after very well'.

And he adds, more seriously, 'Everything I need, every place I have to be at frequently is within fifteen minutes from where I live.'

Indeed many of our interviewees mentioned accessibility of basic services of shopping, schools, and medical facilities as among the most important parameters that determine our well-being in the city. It is what many regard as a crucial advantage of the city over the countryside or small town. It is clearly an urban good. And therefore its distribution among the city's inhabitants is critically important for equality. For example, during the Covid19 pandemic lockdowns there were neighbourhoods which were better served than others. One example reported to us is that delivery drivers, who worked for profit, did not want to go to certain neighbourhoods for reasons of security, or went more often to other neighbourhoods where there was more demand. This created much resentment among those whose neighbourhoods were not properly served.

Another obvious issue is accessibility for those who find it difficult to walk, and for the elderly and those in wheelchairs. Nicky (36) from Oxford, United Kingdom, describes how she adores the museums in Oxford, and yet, she says:

> It's really not great for wheelchair users, which is a pity. Some museums, like the science museum for instance, are non-accessible. And then in other museums, the lifts are too small, they're developed for people on feet but not for wheelchair users. And the colleges here, too; many of them are not accessible.

In thinking about accessibility, it is natural to think perhaps along Maslow's (1943) hierarchy of needs, with services that supply basic needs, such as food and medical services, as the most important. Yet even Maslow had reservations about his hierarchy, and many of our interviewees treated accessibility to leisure services as of enormous importance to their lives, not explicitly relying on any hierarchy of needs. For example, John (30) says of London that he strongly appreciates

> having everything I need so close by my doorstep. Museums, shops, cinemas, restaurants, parks, shopping malls, everything. I'm a city boy. I go to the country for day trips. But you know, here in the city, you never have to travel too far.

Similarly, Yap (67) in Amsterdam says that what he likes about the city is accessibility:

> Amsterdam offers everything one might need: culture, social life, accessibility. Yesterday we went to a small square near our house, where there was live music, and

a nice cafe nearby served food. (. . .) It is much better to age in Amsterdam than in Rotterdam, mainly because here everything is close by.

Such sentiments are widely repeated. For example, Roana, a 37-year-old Londoner describes how in London people have plenty of facilities available within a short distance, accessible by the tube or on foot saying: 'This morning I went to the gym and to Yoga just over there. And I can walk from home. It's very easy to get around here.'

Sumeye (32) from Hamburg highlights easy access to supermarkets, shops, and also amusements for her children as among what she appreciates so much about the city, and indeed nearly all Hamburg interviewees mentioned how easy it is to access green parks and even the waterfront, which remains accessible to all and so the city is called 'Tor zur Welt' (gateway to the world).

Opinions can differ about conditions of access in different cities. In Jerusalem we interviewed Amalia (an orthodox Jewish woman) and Tal (who is secular) together and they had the following exchange regarding accessibility to pubs for orthodox Jewish women:

AMALIA: Jerusalem is the most equal city! As a religious woman I can go to a bar here and feel free.

TAL: Can you? I can offer a different intuition. Jerusalem is a religious place, and so many different complaints from different sectors can be held against a religious girl in a pub. There's much more criticism here.

AMALIA: In my opinion Jerusalem is actually more open to different sectors than other cities. You can see the diversity especially in Nahla'ot [a mixed, religious and secular neighbourhood]. It's not possible everywhere, obviously. There will be no pub in Ramat-Eshkol [an ultra-Orthodox neighbourhood]. But I'm fine with it! If we are talking on a specific ultra-Orthodox neighbourhood, then that's OK, as long as it does not force me to take another road. [By which she refers to rather rare cases when ultra-Orthodox men demand that women avoid walking in public spaces in their neighbourhoods side by side to men, and should, therefore take special, different routes.]

These interviews go hand in hand with the data collected in a survey conducted in 2010 in twenty-six American cities by the Knight Foundation and Gallup (n.d.). Those surveyed were asked to specify the key factors which were crucial for them in order to feel attachment to their cities. In all twenty-six cities social offerings (opportunities for entertainment, social life, places for people to meet each other, vibrant night life, etc.) came first (n.d., Overall Findings, 10). Correlation between social offerings and attachment to the city was much higher than the economy and even safety, not to mention opportunities for civic engagement.

However, for goods and services to be accessible, it is not enough that they are distributed evenly in spatial terms, or that public transport is plentiful; the goods and services need to be affordable as well. Raze (24) came from Geneva to London. After saying that Geneva people are more laid back while people in London seem to either make money or spend money, he does acknowledge that for the average income in London, the cost of living is very high. 'Yes, I never go out on the weekend. I don't have money to do so. I like football and going to the stadium. But that's too expensive here.'

Public amenities, provided free or at very low prices by the city, occupied the thoughts of many of our interviewees. We have already discussed the importance of public toilets, and, also mentioned live music in the park. But the concern spreads much more broadly, and many worry about what the city can offer to children and adolescents. Moti (61) from Tel Aviv complains that although the number of children per family in the south of Tel Aviv (where poorer families reside) is much higher than the number of children per family in the north of the city, there aren't enough playgrounds and very little money is invested in developing facilities for those children in the south of the city. He then comments ironically, 'only soup kitchens for children do they build here'. Channy, aged 36, lives in a gentrified neighbourhood in Tel Aviv, popular with young unmarried hipsters. She was born in the neighbourhood when it was a neglected area:

> There is not even one park here, or one community centre. The young people have bars, but for families and people my age there is nothing. Not even one playground for a mother to play with her children. There are more dog courts than playgrounds for kids.

So accessibility should take into account the preferences and goods sought after by a variety of people, and not only the very young, or the bourgeois, or those who speak up the loudest. And cities should never forget that accessibility is not merely a matter focusing on bicycles rather than cars, or public transport rather than private cars. It should also be about walkability, as Mary Soderstrom, resident of Montreal, Canada, argues. Describing the anguish that she and her husband experienced while looking for a flat to buy, she writes: 'The only thing clear from the start was that we were going to live within walking distance of my husband's work' (Soderstrom 2008, 13) which, indeed, many among today's more fortunate city dwellers find as a major component of their urban well-being. Therefore, sidewalks have to be maintained and fixed regularly; but often municipalities spend more money on fixing them where businesses are located and in neighbourhoods where the more affluent residents live. People literally fall and get hurt more often in areas where sidewalks are not maintained regularly.

Sidewalks and their use, perhaps surprisingly, have long been an arena for political conflicts and debates because some of their uses have proven controversial, from demonstrations to shelters for homeless people (Loukaitos-Sideris and Ehrehfeucht 2012). Sidewalks are where we meet our neighbours for a quick chat, and in some neighbourhoods where children play, but also where cafes put their tables on a sunny day, prima facie 'annexing' public space to their businesses. When a cafe does so it increases the accessibility and enjoyability of its services, but it also can make passing through more difficult for people who are walking past, or, even more so, pushing prams or using mobility devices. The same happens when people gather round to listen to buskers sing in the street. In Copenhagen the city encouraged owners of cafes, and restaurants, but also regular flats and buildings, to put tables and chairs on the sidewalks next to their buildings or businesses, in order to make what they defined as 'edge zones', where the private borders with the public, more inviting for all who pass by (Mclaren and Agyeman 2015, 137). Another regulation in Copenhagen forbade developers of new buildings or people who owned private estates to put a sign 'private' or 'private, no entrance'. This, as the ex-city architect, Tina Saaby explains, literally meant that access to each building became free at all times of the day and that nobody felt she was unwanted or not welcome in a certain building or housing project (Saaby 2015: see minute 15:50). The idea behind these policies is to deliberately blur the boundaries between public and private, which, in turn, should make city dwellers who are less affluent feel more equal and respected.

3.4.2. Frequency and Variety of Public Transportation

Following the themes that relate prominently to the value of non-market accessibility to goods and services, we arrive at transport. The two are actually closely related. Here we turn back to an example that we mentioned in Chapter 1. In Rio de Janeiro we were shown public football pitches that stayed open during the entire night to enable those who work night shifts and want to play before or after work to do so. This policy enables wider access to pitches in a city where football is central to social and cultural life. Alas, when we presented this as an example in a talk in São Paulo, we were told that most working-class people who worked night shifts could not use the pitches because the public transportation system from their neighbourhoods to these pitches was unreliable and often did not work. So yes, the city should have facilities and make them accessible to all, but if public transportation does not work well, many will not be able to make use of these facilities. Still, we would like to say, a city with such facilities, even if for a partial group of working people, is more equal than a city without such facilities.

As we have seen, issues of transportation around the city came up very often, especially as so many city dwellers do not own cars, whether for financial reasons or because the inconvenience of finding parking places and keeping a car well-maintained is simply not worth the effort. Excellent public transport, of course, changes the balance and more people will do without cars if they feel there are ways of getting where they need in good time. Hence the frequency, speed, and comfort of public transport will be critical, and a poor system of public transport can make life very difficult. Admittedly, in some cities there is broad positive consensus (e.g. satisfaction rates with public transit in Zurich is 97 per cent, Vienna 95 per cent, Helsinki 93 per cent (Flash Eurobarometer 2016, 419)) suggesting that the city serves the current residents in an egalitarian way, in that very few feel left out. London residents' satisfaction (86 per cent) is also impressive because of its size, but in general it is difficult to compare cities from different countries in this way because differences might be a function of political cultures, or tendencies to complain which vary in different nations and cultures. However, complaints about unequal accessibility to public transportation were common in the interviews. For example, Dvora, a 60-year-old Tel Aviv city-zen reports with regret:

> You just can't tell someone, 'I'll take the bus and will be there on time'. You can never make it on time. . . . And yet, we depend on this system nevertheless.

Reliability is one important factor of course, as are speed and experience. And the comfort of the journey should not be ignored. For example, until recently London's underground system was designed on the assumption that passengers could easily ascend and descend short flights of stairs, with lifts and escalators only for the longer climbs. It also recognizes in its carriage design that not everyone will be able to get a seat at all times, and standing spaces and handrails are built into carriages. But, of course, as not all are equally able to manage even a short flight of stairs or stand for long, this generates a type of inequality, compounded by the facts that for some people the handrails are too high, and, for all, standing very close to others, especially in the heat, can be unpleasant. Many older people, women, and those travelling with children, prefer to use the bus system even though it is much slower, simply because it's more pleasant. When new lines are built, issues of comfort, accessibility, safety, and convenience are closer to the top of the agenda, but older lines are very hard to retrofit.

In cities that have grown rapidly, residential settlement often out-paces thoughtful public transport, leaving those in less-served parts of the city with daunting commutes. Brazil's major cities are notorious in this respect, with an average daily commuting time reported to be 141 minutes in Rio, and research indicating that from the outlying regions of São Paulo many workers have a commute, standing in a crowded carriage, of an extraordinary three hours in each direction daily. This is part of what is often termed time poverty in the city. Time has become a precious resource for everybody, but especially so for those who have to commute long distances. It is often the case that social position as well as gender can have a bearing on how one uses their time in the city and how challenging time poverty is for them (Walker 2013). Our interviewees from Rio confirmed that this was their experience. Arthur, from Rio reported: 'Life is tiring and stressful and insecure because of poor public transportation in my neighbourhood' and 'the geography of the city keeps people apart from each other.' The lack of transport opportunities can sustain perceived divisions that people already have, often referred to as 'mental maps.' Greenberg, Raanan, and Shoval (2014) interviewed Palestinian, secular Jewish, and ultra-Orthodox Jewish women in Jerusalem to learn about their mental maps, and where they felt 'belonging' and where less so, and then compared it to their actual behaviour, using tracking technology (GPS) and activity diaries to plot the actual use of space. They found a very strong relationship between perceived personal territory and actual spatial activity (Greenberg, Ranaan, and Shoval 2014). We conclude that, in addition to making commuting time reasonable for all, in a city of equals all groups—race, class, gender—go anywhere, or at least can if they wish to, without feeling unwelcome.

From the perspective of inequality, an inadequate system of public transport will impact the poor more heavily than the well to do, for two reasons. First, because those who can afford to own a car or use taxi services will find a solution, annoying and tiring as it may be, to reach their destinations. Second, because public transport systems often leave the neighbourhoods where the poor reside even less well served than more affluent neighbourhoods (see also Soja 2010, vii–xviii). This, we saw, was noted by interviewee Arthur from Rio who suggested that the lines do not reach all the poor neighbourhoods, and where there are buses or trains, there are fewer per day than there are in the more affluent neighbourhoods or in the city centre. This of course, is a self-fuelling mechanism; house and apartment prices will tend to rise faster in areas with good transport.

When public transport does work well, and people from all walks of life tend to use it, there is more mixing of ethnic groups, classes, young and old, and to the extent that there is some exclusion or spatial segregation in the city, it is diminished and becomes less severe. Under these conditions the city feels safer for women and elderly people. In terms of inequality between generations and between the sexes, an excellent public transport system makes the city more welcoming, inclusive, and more of a city of equals.

For this reason, Berlin's former mayor, Klaus Wowereit, attempted to plan that its transport system of U Bahn, S Bahn, trams, and buses, as well as public bicycles and e-scooters, spread all over the city, so that city dwellers can make use of one of the means of transport without walking more than 400 meters from wherever they are. This was considered a very egalitarian move, although in practice the system has not functioned as planned, or at least so it was claimed by many, with the term 'Zug fällt aus' (train is cancelled) an all-too-common refrain. Fatima is 24 and has moved to Berlin recently. She says:

> In many places you really need a car, and you are in trouble if you don't have one. That's where the gap stems from; but this is less pronounced here in Berlin than it is elsewhere.

Indeed, as they also understood very well in Berlin, how easy it is for a person to reach public transportation is not less important than the frequency of buses or trains that serve the neighbourhood. In addition, a question arises as to what would count as equality in access to transportation. Litman (2023) describes the two main answers as horizontal equity (people with similar needs and abilities should be treated equally, so for example trains and buses should serve all destinations with equal frequency) versus vertical equity (disadvantaged groups should receive a greater share of resources, so trains and buses should arrive to their neighbourhoods more frequently).

But even this does not reveal the entire picture. Perhaps the most pressing issue when it comes to transport and inequality in the city is the length of time it takes to travel from home to work or school. In many mega-cities and metropolitan cities in the Global South, this has become a daunting issue, because, as we mentioned in relation to São Paulo, people who live in the distant suburbs can have a commute to work of three hours each way, which, we were told, typically requires waking up at 4 a.m. and returning home in the late evening. Giannotti and Logiodice (2023) developed an interesting way to analyse the inequality of transport in the city: they measure how many jobs can one compete for or apply for, considering commuting time; how many schools can a child reach from their place of residence? With the latter, they found that while prima facie inequality is not substantial; when the quality of these schools is also taken into consideration, it became clear that in some areas of the city they studied—São Paulo—it was impossible to reach a good school without debilitating fatigue, due to commuting time. They applied the same checks for hospitals, clinics, and leisure activities. When they added the cost of transport (those who earn less live far away from all these goods and have to pay even more to get there) they found that inequality was severe. In some cases, people commute four to six hours a day, and they would either work fewer hours in order not to risk the functioning of parenthood (i.e. fulfil their duties as a parent) or work eight hours a day and either risk the functioning of parenthood or sleep very little, thereby risking their mental and physical health.

One theme that came up, especially in European cities, was how accommodating the transport system is for those who cycle, as well as for pedestrians. Although, of course, cycles are a form of private transport rather than public in a strict sense (unless the city purchases bicycles on a mass scale and makes them available throughout the city for the use of city dwellers), encouraging cycling and walking often goes with a public transport ethos, where the main aim is to discourage the wasteful and dominating use of the private car. Cities such as Los Angeles are notoriously difficult to navigate other than by motor vehicle, whereas in Copenhagen the 'green wave' of traffic lights is designed to meet the tempo of cyclists rather than cars. Numerous interviewees were concerned about whether the city is planned for car drivers or for cyclists and pedestrians, with adequate sidewalks and attention to the needs of those who don't have their own car.

Julia (29) from Amsterdam says that one of the advantages of her city is that while it is cosmopolitan, it also feels like a village in the sense that 'everything is accessible by bike.' Interestingly, she points to two advantages that go beyond swift accessibility: first, it is easy to become very familiar with the city, and second, she gets to chat with a lot of people. 'I love the small conversations I'm always having just cycling down the street. These small interactions are the best.' Interestingly, Julia's positive feeling extends beyond considerations of the convenience and well-being of cyclists but includes a sense of togetherness that people have when so many of them ride bikes to and from work or study, shopping, visiting friends, and picking up their children from their schools. The city is designed in a cycle-friendly manner and is tolerant to them, and it just feels good, as well as safe, to be part of a community that travels by bike.

This testimony concurs with a claim made by Tina Saaby (2015). Describing Copenhagen's *bikestrategy*, which was part of the city's goal to make people spend 20 per cent more time outside their homes in public spaces, she explains how the city's urban planners made it more time-consuming to use private cars, hoping to incentivize people to ride their bikes more; the result was that this indeed happened, and people described how they enjoyed the experience of talking to other cyclists while riding their bikes in the bike lanes.

3.4.3. Gardening, Environment Goods, and Environmental Bads

In the past, cities were often surrounded by walls which were meant to protect the city not only from human animals, but also from wild beasts. Nature was perceived as distinguished from the city. Researchers today call this attitude the city–nature dualism thesis. As Barak explains:

> The relationship [was seen] as oppositional and antithetical—the city is not conceived of as being part of nature and vice versa. Dualism incorporates two standpoints—one which regards cities as 'bounded social containers' in a nonsocial nature and another which sees urban parks, wildlife, etc. as islands of 'nature in cities'.
>
> **(Barak 2020, 56)**

But, as Barak continues to explain, most contemporary urban planners and environmental activists subscribe to the view that this thesis is false, or at least should not guide us when we plan our cities. The bird that nests on a tree in the street we live in or on the roof of one of the city's buildings, is part of nature and of city at one and the same time. And so are city-zens. National Geographic's website recommends several ways to enjoy what they call *urban nature*: volunteer at a community garden; pay attention to the way the city looks in different seasons; on stormy days or if it's too late to go outdoors, experience nature by watching webcams of birds of every kind; plant a tree; look out for flowers in your neighbourhood; go for an early day walk, before the day's bustle begins, to watch the animals you can see around you; collect leaves, and so on (National Geographic n.d.).[3] The intended

[3] We do not here discuss the relationships between humans and non-humans in the city. For an interesting discussion of several questions with regard to the interplay of cities and the non-human world, see Epting (2023).

audience is schoolchildren, but many adults too would benefit from these activities. The urban nature atlas describes more than a thousand nature-based solutions to urban environmental problems. For example, in October 2021 in Boston, United States, the Barr Foundation started a programme called Waterfront Initiative, which has two goals, environmental and civic: 'to support the sustainable planning and accessibility of the waterfront area while also increasing residents' awareness and environmental stewardship of the waterfront' (Urban Nature Atlas 2023). Scholars emphasize the health benefits of urban nature (Shanahan et al. 2015) which includes physical health as well as cognitive and psychological health.

Indeed, from our interviewees we learnt that city-zens appreciate the many aspects of nature in the city and they think that being deprived of access to urban nature—to the extent that this happens—is a key component of inequality in the city. Amar (65) immigrated to Amsterdam from Tunis thirty-eight years ago. He sees access to parks as a comfortable and less unpleasant way of aging in the city:

> I wake up, take a stroll in the park. (. . .) Sometimes friends are here too and we get together. Then I just walk around a bit, having some beer on the way, and by 8 pm I'm home.

Brenda (60) migrated to Rotterdam from Romania. She also believes that access to parks is especially important for the elderly: 'There are many parks here, and an old woman like me enjoys feeding the ducks.' Women also mention how they often find it difficult to enjoy the park as an issue of gender inequality. Maria is a relatively young Jerusalemite, who complains that she has to avoid the Sacher Park, close to her home, because of harassment. Moti, in Tel Aviv, complains about inequality between the haves and the have-nots, because the latter reside in the south of Tel Aviv 'where you do not see parks.' Chany, who also lives in the south of Tel Aviv, is asked 'what comes to mind when you think of inequality in the city', and answers immediately, referring to her own, less-affluent neighbourhood: 'There is not a single park here' Tamir (40) from Tel Aviv, describes a different situation in the north of the city: 'I really grew up in and with the park; I was there all the time.' Mariza from Rio de Janeiro mentions especially Parque Madureira, describing the fun of chatting with others in the open area, and Renata, also from Rio, claims that 'there are fewer parks and nice open spaces for citizens in the North and West Zones [where the poor live] to enjoy during their free time.' Arian (38, male) from Oxford, lives in East Oxford, but visits regularly the various parks and meadows, as well as the college gardens: 'What I like about the city? There is lots of green (. . .) the park here, but also the meadows, and then obviously there's all the colleges with their own greens. I really like that.'

We have argued above that the importance of public spaces in general is not simply that they are out there, but that they are accessible and usable, and that people of all races, ethnic groups, ages, attributes, and from all neighbourhoods, should feel welcome in them. People want to feel that they can enjoy urban nature according to their current needs: if I come to a park alone, I can sit to rest on a bench, and it does not matter very much how the bench is arranged as long as it's in a pleasant spot. But if I come with six other friends, we'll probably need two or three benches facing each other. In his widely read *The Social Life of Small Urban Spaces*, William Whyte (1980) examined the most important factor for determining whether people will or will not use public space in the city and found that it was a combination of (a) enough shade, and (b) whether there were accessible chairs or benches and whether they were movable and attractive to use. Indeed, Amanda Burden, who was NYC's chief planner when Bloomberg was mayor, explains why people near Paley Park, a pocket park in Manhattan, are very fond of visiting the park on a daily basis. The park had a profound impact on New Yorkers, she claims. One of the reasons was comfortable, movable chairs. People would come, find their seats, and move them a bit, to suit the way they wanted to sit (Burden 2021). We conclude that parks in egalitarian cities should be planned so that they can cater to a variety of groups; those who visit the park in large groups or families, those who visit the park by themselves, and so on. They should also be open as long as possible, and never carry the message that some people are not wanted.

Perhaps even better, and as Jane Jacobs taught us, to cater to the needs of recreation and playing outdoors for all, cities should design small parks in each neighbourhood or at the end of the street, in addition to spacious parks such as Hyde Park or Regent Park in London, Tiergarten in Berlin, Bois de Boulogne in Paris, Century Park in Shanghai, Flamengo Park in Rio de Janeiro, or Central Park in NYC, to mention just a few of the world's most famous parks. These huge parks are of course impressive and attract many locals as well as tourists; but most city-zens have to travel by train, bus, or car in order to reach them, and so accessibility to these parks becomes an issue. In addition, as Jacobs suggests, small, human-scale gardens should be planned around the corner.

But some, wealthier, city-zens pay more local taxes than others, and some—very often the same people—have better access to decision-makers. The result is that sometimes, to return to the topic of trees and shade mentioned above, city authorities tend to plant trees in neighbourhoods where municipal tax receipts are high, and to avoid planting where for socio-economic reasons residents do not have to pay, or pay significantly lower, taxes. This produces heat islands, and different micro-climates, which can make an enormous difference during heatwaves. Heat islands can also be part of the explanation, alongside the lack of local social and material resources to cope with extreme heat, why in certain city neighbourhoods heat waves cause deaths whereas in other neighbourhoods in the same city there are far fewer fatalities (Davis 1997; Klinenberg 2002; Harlena et al. 2006).

Access to gardens and parks should ideally include private outdoor spaces, such as allotments to grow your own vegetables, although of course this is not practical in very densely populated cities. But where allotments are available, they provide a hobby but sometimes also occupational therapy, or simply extra food for the family. At the moment it is often the case that the more affluent who want or need such practices can rent allotments outside the city and drive there. If cities can provide more allotments near every neighbourhood there would be many beneficial effects including the reduction of inequality in the city.

But when it comes to the environment, cities distribute not only access to parks and gardens, but also exposure to environmental bads, which, if they cannot be avoided altogether, should normally be distributed on an equitable basis, although there are exceptions.[4] There are at least two kinds of exposure to environmental bads, involuntary and voluntary. Naturally, the involuntary is the more problematic. By this we mean that the more disadvantaged city-zens find themselves residing by the sources of pollution, or the less beautiful quarters, while the rich enjoy calm views, and a well-kept and quiet environment where they do not have to face industrial pollution. So, for example, Petra (60) from Berlin says you can see how different neighbourhoods are treated if you look at the cleanness of their local train stations. Explaining why she recently moved to the neighbourhood where she lives, she says something many among her age would appreciate: '[This is where] I can immediately

[4] How are environmental bads (pollution, noise, garbage) distributed? Some of these bads are associated by many with cultures; namely that some cultures tend to care less about littering in the public domain, or about making noise. Nona (59) complains that in the neighbourhood where she lives in Jerusalem there are many ultra-Orthodox people who are mostly poor, and, she claims, therefore the area is always filthy. But unlike others, who blame the city authorities for not cleaning up, she believes that it is a cultural thing, as 'they have their own ways'. However, we want to note that littering might be a cultural thing, but it is surely also structural: if local facilities are poor, with no garbage cans outside, people might be tempted to litter. Even more problematic is irregular emptying of public garbage cans, which can overflow or be disturbed by birds or animals spreading litter on the streets. Even if everyone has disposed of their litter responsibly the streets may still end up in a very poor state. See, for example, Schultz et al. (2013), and Carpenter (2014).

relax—and that's something where noise, or rather its absence plays a big role.' Dvora (60, Tel Aviv) complains that she pays taxes, but in her neighbourhood the municipality does not clean often, as it is a neighbourhood occupied by less-affluent residents: 'I don't get anything back! Nothing! They don't clean here, and there are even no public toilets here.'

The other kind of exposure to environmental bads—the 'voluntary' form of exposure—is where those living in the city centre are exposed to noise and pollution, but that is a price they are ready to pay. Several examples for the latter come to mind: St. Pauli in Hamburg, Soho and Ladbroke Grove in London, 5th Ave, or by Central Park in NYC, Central District in Hong Kong, and many more. Sivan from Tel Aviv lives in a trendy neighbourhood with lots of bars and cafes. She says:

> Obviously the neighbourhood is very noisy. But I live the noise [i.e. it gives me life], I need it. I can't be surrounded with quiet; I need to hear some life around me. I really can't have the quiet.

Eddi (32), also from Tel Aviv, says: 'You know, for some time this kind of noise is actually very pleasant.' It is often a matter of age, but Hanna (70) who used to live in Jerusalem and moved to Tel Aviv says that noise shouldn't be a problem. 'I actually like it,' she says. It seems that David (50) from Tel Aviv is right when he sums it up: 'Some people like quietness; I don't. People are attracted to different places because of who they are.'

Some interviewees suggested that environmental matters impact the way people behave to each other, as when we mistreat the environment we tend to mistreat each other. Moti (61) from Tel Aviv says 'it is all cement around; it is so symbolic. (…) People are wolves to each other (…) The city has become like that; we are all capitalist pigs.' However, when Jane Jacobs was writing, American planning was turning to zoning, and factories were being moved out of town centres. This, many urban historians claim, killed communities (Silver 1997; Gray 2022a; 2022b) and in that sense was not sustainable. Therefore, we wish to note that of course pollution and noise are, at least in physical health terms, bad and exposure to them is often risky and harmful, but physical health should not be the only consideration. Some light industry could be beneficial in terms of equality, in bringing different types of people to the same neighbourhood, who will then have to find ways of peaceful coexistence. It is almost a paradox that an egalitarian city may well have more pollution than a tranquil, upper-class, city. This is because the egalitarian city will want to create a welcoming and encouraging environment for poorer people and allow a variety of opportunities for people to flourish and get out of poverty (Alster 2022). So it probably needs jobs in industry as well as people on the move, using trucks, older cars, and so on. The result is that this egalitarian policy will inevitably create more air pollution.

Michael Young and Peter Willmott (2011 [1957]), in their classic study *Family and Kinship in East London* looked at patterns of housing and work in London. In the 1950s families who had been living in the slum areas which had been badly damaged during the Blitz were re-housed to new towns in the suburbs. For the men this was largely a positive experience because their improved housing did not stop them from continuing to work as they had done before and thereby keep their social ties, for example, going for drinks after work. But their wives had a miserable time because they were taken away from their neighbourhoods. Women in the study cared more about contact with friends and close family than about improved housing. The book shows that although crime dropped and husbands spent more time at home with families than they did before, overall there was a serious harm to the community. In our book *Disadvantage* (Wolff and de-Shalit 2007) we describe a similar process. As we have already mentioned, John Bird, the founding editor of the magazine *The Big Issue* told us how when the city wanted to help him and others in their run-down neighbourhood, a decision was taken to rebuild, but because residents were dislocated in order to enable the construction, they lost their communal ties and the relationships which had kept them all functioning despite their poverty, to the extent that many of them fell into crime, drug addiction, and so on.

Nevertheless, we cannot avoid the fact that exposure to environmental bads is harmful, as it is seen in any theory of environmental justice. Moreover, it is not only a matter of physical health. In addition, exposure to environmental bads is often described in terms of exposure to the stress of the 'hustle and bustle' of the city. Mano (70) from Jerusalem enjoys the tranquillity and quietness where he lives and says that at his age this is an escape from the noise and chaos of the city. Neusa is 59 years old, an African-Brazilian woman, born and raised in Rio de Janeiro. She picks through garbage for aluminium cans to sell to recycling companies. She lives in both a violent and filthy area of the city together with her son and four grandchildren. She says all she longs for is to live in a 'quiet environment', referring both to the violence and the hassle and noise around her. But Neusa is disadvantaged in her city not only because she is exposed to hazardous waste, violence, noise, and the hustle and bustle of the city. She is also homeless. Which brings us to one of the main themes for a city of equals, as described by the interviewees.

3.4.4. Housing Policy

Housing was an enormous concern for our interviewees, and with regard to several related issues: housing availability and affordability, most often discussed in terms of rent, which has been rising rapidly in many cities in the early 2020s, especially as a percentage of average gross pay; the proportion of households in temporary accommodation; chances of owning a flat in the city, especially during times of rapid house-price inflation; and also the number of people seen sleeping rough.

Regeneration and urban renewal, discussed under the heading 'gentrification' for most interviewees, was a common topic. While many interviewees appreciated the facilities that regeneration brings, they were concerned that the process spirals to attract business or wealthier newcomers and drive out long-standing residents. Erika, 75, from Hamburg says:

> They are tearing down buildings. Buildings that are not too bad actually. And then they're constructing these huge and really expensive office buildings, and no one can pay the rent.

Klaus from Hamburg says:

> The rents! We really need to regulate this, it's completely out of control! Where are all these people supposed to live? We'll end up with a drastic situation where those who have money can afford to live in the city, and all others will have to leave. That's unacceptable.

Nearly all interviewees in Berlin told us that housing prices are too high. Similarly, Sandra (49) a designer from San Francisco, now in Berlin, reported that she had moved to San Francisco because it was a walkable, bicycle-friendly city and it had a European atmosphere, which she liked. But then the city became very expensive. Initially teachers and artists had to move out, then more and more people found it unaffordable. Tina from Hamburg also complained about the housing prices:

> Up till now Altona North [a neighbourhood in Hamburg] was a place also of the 'simple people', normal people, and by that I mean the medical nurse and the truck-driver. But now they can't afford it anymore. Real estate, housing, it's all too expensive.

She is not the only interviewee who also subscribes to the view that cities are being damaged by investors treating housing as a financial asset rather than a home. It was not unusual to suggest that cities will become much more egalitarian if they introduce regulations to prevent or at least reduce foreign investments in housing:

> I'd introduce stricter regulations for the real estate market, so that investors from China or the US cannot buy property here as capital investments. If you take a look at the houses around the Alster [the main river of Hamburg] at night every second house is empty. I'd prevent this from continuing.

Another interviewee also called Tina, who is 74 and from Oxford, said house prices were the most inegalitarian aspect of the city.

> I think the big problem in Oxford is the cost of the housing, that is a big factor. So people on middle incomes . . . nurses, teachers, those essential industries . . . find it very hard to find anywhere to live in the city.

Many interviewees were aware of the benefits of gentrification as well, namely that neighbourhoods become vivid and full of life again, where otherwise they would have continuously deteriorated, especially in areas where cuts to the budgets of local authorities make it very difficult to intervene by using public money to save run-down areas. However, while from the perspective of the city regeneration or urban renewal private investment is welcome, current residents who are forced to move out are understandably very bitter about it, and their plight evokes wide sympathy. N (33), from London, answers straightforwardly when asked what comes to mind when he thinks about inequality in the city: 'Housing!' He waits a bit and then elaborates:

> There's many landlords buying property massively. And as a consequence, people are being forced out. Gentrification is happening. I know in Germany they have price-capping, so maybe that's an idea. (. . .) OK, I have to choose my words carefully here. I think in many poor areas, socially weak, perhaps not so safe . . . when new people move in, they can improve the neighbourhood . . . they upgrade it, and the situation as a whole. (. . .) But some people [are particularly vulnerable], those from a lower socio-economic background. That is not necessarily Black and Asian communities, I think . . . it's not necessarily race.

Following rents, house prices, and gentrification, interviewees associated inequality in the city with the connected issue of the low supply of affordable housing, either publicly owned or by private entrepreneurs. Maaike (60, Amsterdam) believes that the worst thing about inequality in Amsterdam is that it has too little social housing; affordable accommodation for people who can't afford the market rent to live in the city. The 'ridiculously long waiting period for social rent housing motivates people to leave the city'.

It is interesting to note that Maaike's comment runs against one of the leading academic studies of justice in the city. In her famous discussion, Fainstein argues that Amsterdam's housing policies were responsible for keeping the city relatively egalitarian. Housing policy has been a major instrument in maintaining the quality of life for the city's lower-income population. Because subsidized housing units, as well as recipients of individual housing benefits, are scattered throughout the city, housing policy has sharply restricted spatial inequality of households by income. Moreover, the very large public subsidy involved in housing construction, by keeping rent levels low and thereby raising disposable income, has contributed substantially to popular welfare, mitigating class differentiation and thereby weakening resistance to residential integration of different income groups.

It is worth reflecting on the difference between Maaike's perspective and Fainstein's analysis. Maaike is aware of the many regulations that the city has in place, and which Fainstein applauds, but nevertheless believes that more has to be done. And this may go hand in hand with the increasing salience housing has had for people in the last decade, with increasing pressure on authorities to treat the matter with urgency in urban settings. For example, The Eurobarometer survey asked what are the three key urgent topics for the city authorities to face, and it shows that already in 2015 housing was mentioned many more times and by many more people than several years before. For example, in 2015 it was more likely to be mentioned as an important issue than in 2012 in Dublin (45 per cent, an increase of 25 per cent), with large increases (of at least 10 points) also seen in five other cities (Flash Eurobarometer 2016, 165). Fainstein's book was published in 2010, so it might be even in this short time the shortage of affordable housing has become much more important in cities around the world.

But there are also factors more specific to Amsterdam, and in particular related also to the change in its the local government. When Fainstein was doing her research Job Cohen, of the Labour party, was mayor of Amsterdam. He was known for his slogan 'keeping things together', emphasizing his attitude to the variety of ethnic groups and the city's responsibility for all its residents. Our interviews were taken during the period in which two mayors served, van der Burg and van Aarsten, both from the 'People's Party for Freedom and Democracy', a centre-liberal-conservative party. Things changed since Fainstein conducted her research.[5] Affordable housing was a key topic

[5] Already in 2009 a paper by Justus Uitermark (2009) suggested that although Amsterdam had become a just city it was then dying. Uitermark claims that both equitable distribution of scarce resources and democratic engagement, two essential preconditions for the realization of a just city, were disappearing from Amsterdam's scene.

in the March 2022 local elections, and a recent housing survey conducted by the municipality (Sevano 2022) found that, between 2019 and 2021, the number of privately owned rental properties increased by nearly 10 per cent, to over 137,000.

More recently, under the leadership of the Green mayor Femke Halsema, new regulations have been announced, intended to mitigate scarcity by preventing flats in the city from remaining unoccupied. In 2019 the mayor declared that homes are meant to be lived in, not to earn money (Halsema n.d.) explaining why she was in favour of limiting the time a flat can be vacant. But whatever the reason for these differences, the relevant, though unsurprising, conclusion for our argument here is that affordable housing in particular, and housing in general, is perceived as a key element of a city of equals.

In moving forward progressively, the egalitarian city can be innovative in its solutions, and this is an area where new experiments (literally 'experiments in living') would be very welcome. The American Community Development Corporation (CDC) is a good example. CDCs are vehicles for supplying affordable housing, below market prices, in the United States by subsidizing tax reduction and philanthropy (Bratt and Rohe 2007). In most CDCs there is representation of the residents on the governing board. A good example is Dudley Street Neighborhood Initiative in Boston (DSNI n.d.). Not only does it lead 'development without displacement', namely a process of development and urban renewal in the neighbourhood that ensures that the neighbourhood's residents are not displaced and that it can persist and flourish, but it does so by means of empowering the local residents so that a long-term effect is much more likely.

3.5. Themes that Relate to Sense of Meaning

We now come to two themes which are closely related to having a sense of meaning, or of a meaningful life. These are, first, how people experience the city, the urban public space, and the city's amenities; and second, having a sense of security, especially in the public space, including having a sense of identity which in the city can be simultaneously communal and individualistic.

3.5.1. Inequalities in Urban Experience

When asked what they like about living in the city, many people answer in ways that draw on the particular character of urban experience. In the beginning of the twentieth century Simmel (1903) observed that the city with its stimuli and rush offers us anonymity. However, today many urbanites would suggest that emphasizing the anonymity the city offers is a very one-sided picture and would argue that the city can offer a very pleasant sense of community life. The city is full of action, sources of enjoyment and opportunities, and huge variety day by day. Urbanism, the urban way of life, which for Simmel meant anonymity, is thought by city dwellers to offer joy, satisfaction, and even self-fulfilment and a meaningful life.

But this is not experienced evenly by all city dwellers. Jerusalem is thought by many to be a very beautiful city. When the British ruled in Palestine (1917–1948) they put regulations in place to maintain the tradition of building in local stone (rather than cement) for aesthetic reasons. As stone became more and more expensive, the regulation changed, after the formation of the state of Israel in 1948, so that the walls had only to be covered by stone veneers which are cheaper. Still, the city managed to retain its unique atmosphere and beauty. Many city dwellers like to stroll in different parts of the city in different hours of the day, as they do in many cities. Yet the sense of beauty is not enough to make all feel calm and welcome. Jerusalem, is, of course, a bi-national city of Arabs and Jews, and many Arabs from the east city feel that they live under occupation,[6] and for the Arab population strolling in different parts of the city during all times of the day is not possible, or will at least result in feeling out of place. Arabs in Jewish parts of the city very late at night or before dawn in the morning are likely to be confronted by police patrols and even if not, might experience a sense of hostility when they meet others. Such fear, perhaps on both sides, creates a division whereby those lacking the confidence born of superior political power feel they'd better stay in their neighbourhoods in the eastern part of the city during these hours. So Jews and Arabs in Jerusalem experience the city in very different, and unequal, ways.

In Rio de Janeiro, 10,302 kilometers from Jerusalem, two interviewees tell Katarina, our research assistant, completely opposite stories about how they experience their city. Neusa, a poor woman who was found begging, keeps saying that what she dislikes about Rio is that she is never safe, her grandchildren are not safe, and there is always violence. Mario, a professor at a local university says that what he likes about Rio is what he describes as being able to 'walk *freely* around the city'. The point here is not merely that

[6] In the 1967 Six Days War the eastern part of the city, which had been Jordanian, was occupied by the Israeli forces. A few weeks later part of the eastern city was annexed to the state of Israel. Unfortunately, the Arab residents were not granted full citizenship and many also would not want Israeli citizenship, as they identify as Palestinians.

different people have different *subjective* notions of what they see and we are not promoting a purely relativist theory of how the city is experienced. People experience the city differently because their lives objectively differ, because their environments differ, and because their urban experiences vary, in addition to any purely personal or subjective factors, such as their general disposition.

Mario himself is not a relativist, nor is he naive. He continues, sadly describing 'the symbolic and physical divisions'. He says:

> Can you believe that either by observing the way a person talks, wears her clothes, or in the way she walks or by the colour of her skin, I can identify the zone in the city which this person comes from?

Similarly, we ask city-zens of Rio to describe what they like and dislike in the city, and Quésia, a homeless woman, reports that she lives in constant fear of paedophiles who, she says, might 'steal my daughter', whereas Renata, a young psychologist asserts that what she dislikes about the city is that 'I have seen a lot of holes in the sidewalks, dirty streets, and bus stops that don't work properly. This makes our life difficult.' It is clear from these interviews that obviously, in Rio, people experience the city very differently, as a function of their socio-economic status. When Katarina asks Renata whether she sees things that might make life difficult for others, she points to the many social inequalities. When Katarina asks again, about how all the inequalities she described affect her, she answered frankly, 'I think I got used to seeing this inequality, and I try to protect myself.'

Division is experienced in different ways in different cities, including in some of the wealthiest cities in the world. 'The Village' in NYC is a pleasant place to stroll. Or so many people believe, because they are not out on the streets very early in the morning. Between 6.00 a.m. and 7.00 a.m. there are dozens of people cleaning the streets and the amount of filth and garbage they have to remove is just unbelievable. Many tourists and locals like walking in the Village because the streets are cosy, charming, and clean. The atmosphere is pleasant, generating a feeling of security. It is said that Jane Jacobs was inspired by the Village when she wrote *The Death and Life of Great American Cities*, the Bible of many city planners even today. But the Village before 7.00 a.m. is seriously unpleasant. In some streets it is almost impossible to avoid stepping in garbage, which is spread everywhere. We walked these streets in those hours deliberately and it generated a complex set of feelings. One is sheer incredulity about the thoughtlessness of those littering to such a degree, and their disregard of others who have to walk through their garbage. Another feeling is contempt that even today many have the entitled expectation that others will clean up after them. Overall, we felt a kind of both personal and vicarious humiliation that other human beings are prepared to act in this way.

Experiencing the same neighbourhoods in different ways is a form of inequality in the city. Although there is a divide between those who clean versus those who walk in the clean environment, that is not our main concern. Obviously, those who clean do not enjoy spending time in the filth, but it is reasonable to claim that at least they are being paid to do so, and all jobs come with burdens. Rather we have in mind those whose work in sectors with unsocial hours or require a very early start and have to walk through still-filthy streets on their way to their work, in cafes, or cooking breakfasts in diners, or as a cleaner or concierge in offices, apartment buildings, or hotels. We attempted to interview some of these workers very early in the morning. Many, understandably, apologized that they were in a hurry, but some stopped to speak. Those we spoke to had typically left Harlem, Queens, and the Bronx between 5.30 a.m. to 6 a.m. that morning, in order to be on time and open the cafe or arrive for their shift in the hotel. All of them told us they rarely were able to make use of Manhattan's entertainment because they lived far away, it is expensive, they have a family they want to spend time with, and so on. And yet, every day they have to walk through the garbage left there by those who had a good time the night before. It is a very stark form of inequality in the city.

Kenneth Galbraith (1992) introduced the concept of what he calls 'the functional underclass' who will include many of those people who in contemporary cities are compelled to face walking through the filthy streets in the early hours. Galbraith suggests that others do not even see or notice members of the functional underclass, partly because they go to work when most of us are still at home, asleep, or just waking up, but also because they are not noticed even when they are there. But our point here is not that early morning workers have to leave home early but rather that in some parts of big cities they experience walking in public space very differently from other city dwellers who in this respect and many others are more privileged.

Other researchers have noted that how we experience public space is of critical importance to how we think of our place in the city. Sharon Zukin (1995, 42–3) argues that through 'mingling with strangers' in public space a shared urban citizenship (our notion of city-zenship) and a shared public culture is constructed. Zukin claims that in such places of meaningful public culture one can find civility, security, tact, and trust. We add that it happens spontaneously rather than through discussion and planning. People feel it, they sense it, they know it when it happens. And when it does not happen, people can sense a form of urban alienation.[7]

In the 1950s, Galbraith (1958) argued that while many middle- and upper-class people could purchase all kinds of goods, among them some luxury items, our public space is dirty, polluted, and unsafe. Since then much has improved with regard to safety, but with regard to dirt, as we have noted, while the city can look clean at least during office hours, the effort needed to make that happen is immense and those using the city at other times of day can have a very different experience. This is part of what makes some jobs in the city less appealing than others.

Galbraith also claimed that in the affluent society people care more about having an enjoyable job than about how many hours they work. It seems that in cities jobs differ in terms of the experience they offer, and this is another respect in which the distribution of work makes up part of the picture of inequality in the city. More enjoyable jobs in general also tend to be better paid and also allow more leisure time, and the resources to make good use of it, although some very highly paid jobs leave very little spare time (Markovits 2019). In theory non-enjoyable jobs appear at least to allow some leisure time, as they are usually eight-hour shifts, whereas those in more enjoyable jobs may even put in ten and twelve hours a day, as Markovits notes. But as things work out, the non-enjoyable jobs typically do not pay well, and since rent in the city is so high many workers have more than a single job, and the exhaustion of work means that their scant leisure time is used for rest rather than enjoyable experience. In addition, as we described earlier, in some mega cities, commuting to work takes so long it leaves no leisure time at all.

As Michael Walzer argues (1983) societies distribute access to leisure time; but in cities this is even more important, partly because when asked about why they like living in the city, many city dwellers refer to the various opportunities they have to enjoy their time away from work. But making good use of available time can be even more important for those who are retired, where stark inequalities in access to cultural events is apparent. Klaus, 73 years old from Hamburg, praises his city for this:

> I like going to cultural events; not only high culture. I go to concerts and performances in local schools, but also to the State Opera, the Elbphilharmonie . . . Not every week, you know, but I appreciate the offers and opportunities. That's one of the big advantages of living in such a large, affluent city.

[7] Leonard Cohen's song 'Please Don't Pass Me By' comes to mind. Cohen describes NYC at the time living there was known to be harsh for many. He describes how he was walking in NYC and brushed up against a man in front of him. Cohen felt a cardboard placard on the man's back. When he passed a streetlight, he could read it. It said: 'Please don't pass me by—I am blind, but you can see—I've been blinded totally—Please don't pass me by.'

But, in contrast Ruth, 89 years old, also from Hamburg says:

> If it weren't for my son, who is financially supporting me, I couldn't even pay rent. So yes, I'd love to go to the theatre and to the opera as well as to other cultural events, but I can't afford it.

Lack of access is also a problem, of course, for those still of working age, as we noted above. Christian, aged 40 from Berlin, makes the point when he says that he feels excluded from a 'decent life': 'Without a very good salary you can't really afford much in Berlin. Something should be done about this, really.' To our question what is included in his idea of a decent life from which he feels excluded, he says:

> Well, cultural offerings, events, for example. You know, being able to participate [he used the somewhat stronger term 'teilhaben'—which in a more literal translation would mean both participation and co-ownership]—I think if you like to participate in cultural events—in order for you to be able to do so . . . now, well, you have to pay for it. Also exhibitions etc., it all costs money of course.

In other words, accessibility to cultural events, which is part of the urban experience, requires not only availability but also affordability for city dwellers.

Many city residents, in the most populated metropolitan cities, especially younger workers new to the city, live alone in a tiny 'box' of a single room somewhere, or have a room in an apartment, often shared with people they had not met before. They work hard, and a peak experience of leisure time is when they meet their mates and colleagues for a chat and beer after work. The city enables us to really enjoy these moments, as there are so many opportunities to dine out, sit in a cafe, stroll in the park, or sit in the pub, or more often now, standing outside the pub with friends. A famous jazz song describes NYC as great fun because you don't need to spend a cent on having fun—all you do is sit in the park and watch people go by. This is true; but one needs free time and enough energy to spare even to do this. Those interviewees who had to start their work at 6.00 a.m. or 7.00 a.m. reported that they rarely enjoyed leisure time in this sense.

There is something else you need in order to enjoy the rich variety that the city offers: an attitude that others in the urban space express to you that makes you feel welcome to use and enjoy these attractions. Margaret Kohn (2011) notes that formal and informal rules about the use of public space can lead to unintended patterns of class segregation, which, viewed from the point of view of the tastes and interests of each individual could enhance everyone's personal experience, yet seems problematic from the point of view of equality. Kohn discusses the work of Frederick Law Olmsted, the nineteenth-century American landscape architect who is today considered by many to be the 'founding father' of landscape architecture. One of his most famous projects is Central Park in Manhattan, NYC. Being sensitive to inequality in cities, Olmsted hoped that urban parks could enable people from different walks of life and different neighbourhoods to mix and mingle, which, he thought, would enable them to also feel more empathy to each other, and then perhaps support city measures to reduce inequality. However, Kohn argues, Olmsted did not pay sufficient attention to how formal rules and informal norms and customs can make the use of parks more fitting to the tastes of bourgeois residents and make poorer and working-class people feel uneasy and unwelcome there. Olmsted, like many of the middle-class New Yorkers who used the park, regarded it as a work of art, and expected visitors to treat it as such. Therefore, the use of the park was regulated by strict codes, 'similar to the rules governing military recruits' (Kohn 2011, 83). Playing musical instruments, for example, was forbidden in Central Park, together with fishing, picking flowers, and many other activities. In practice the rules enforced middle-class conceptions of proper behaviour, and poor and working-class people did not feel at home at all. Instead, they favoured Jones Woods Park, that 'allowed more boisterous recreation, including spectacles, beer tents, games of chance, popular music, competitive sports, dancing, and large picnics' (Kohn 2011, 84; Rosenzweig and Blackmar, 1992, 233). The use of space is critically important for opportunity and well-being, and this is something worth reflecting on, alongside Kohn's discussion of Central Park. It is tempting to think that a city of equals should offer everyone a similar experience. Yet it may be impossible to find a similar experience that suits all tastes and interests. Hence significant diversity in a city is not always a problem from the point of view of equality so long as everyone feels included in the city in their own way. This is important, for if people feel strongly identified with one group and one neighbourhood, but do not feel embedded in the city as a whole, then a critical aspect of a city of equals is missing, even if everyone is contented in their own way.

Before we move on to the next theme, we want to comment about access to places of worship such as churches, synagogues, and mosques. We were slightly surprised that not many interviewees mentioned them as part of what constitutes their sense of meaning in the city. Obviously, for those with religious faith, access to places of worship will be critical to their sense of being able to live a meaningful life. It might simply be that the issue didn't come up because everyone who values such facilities is already well served; perhaps in the cities where we conducted interviews there is no shortage of such buildings and facilities, so people do not think of it as a matter of inequality. Still, we argue that because there are diverse religious populations in all cities, having a place to pray and worship your God or Gods is also a crucial part of a city of equals.

3.5.2. Sense of Security

According to Flash Eurobarometer 419 (2016), Jane Jacobs was right: there is a very high correlation between satisfaction with one's city and feeling safe, both in one's neighbourhood and one's city.

Indeed, security or safety, or lack of, is an essential element of city life. In fact, it is mentioned in many interviews that we have already discussed, particularly where we consider who is affected by violence and who benefits from policing. It is also mentioned in the context of gender, age, childhood, violence, harassment, and ageism, all of which we discuss below.

One question which is revealed in these interviews is whether what matters is that people *feel* secure or *are* secure. The former is about a subjective notion of security, whereas the latter is about an objective one. Many interviewees talked about both, which is consistent with work we have done exploring the difference between objective risk and subjective fear, noting, for example, that one way to reduce objective risk is to increase subjective fear, so people take higher levels of precautions (and vice versa) thereby lowering their risk (Wolff 2006). What was especially interesting, however, is that in our interviews city dwellers often tended to understand security where they live, in their neighbourhood, in objective terms: they refer to widely available knowledge, including rough statistics, about crime, violence, and sexual harassment. But they also think about security in parts of the city with which they are less familiar, and they do so in subjective terms, or they rely on what might be misconceptions of the level of security in these neighbourhoods, especially neighbourhoods where many immigrants reside. Accordingly, many interviewees felt much less safe when they strayed from the familiar; a feeling which is closely connected to a sense of not belonging or being a stranger, or of being watched, even if this is empirically baseless. Outside their comfort zone, where they live, our interviewees tended to rely less on well-founded information about how secure this area is, and more on scare stories that they hear from the press or as rumours. This was confirmed in interviews in Rio de Janeiro. Even people who resided in neighbourhoods that objectively are very unsafe, for example, in the favelas, expressed anxiety and fear about going to other city zones, which were statistically much safer. What we observed in other studies is that when people need to experience something as part of their daily life, such as regular commuting, they tend to downplay the risks, while those for whom the activities are optional and less frequently encountered are more likely to be swayed by exaggerated scare stories (Wolff 2006). Similar reasoning can apply here: one cannot avoid one's own neighbourhood, but it is much easier to stay away from others. While the explanations we have offered here might seem as nothing more than hypotheses, the important bottom line is this: it seems to us that both objective and subjective risk matter in that we want people to have a feeling of security and we want them to have a good reason to have this feeling.

Yet how a feeling of safety is achieved is not a simple matter, as we have already noted. For example, and particularly relevant to our topic here, there is a complex relationship between the feeling of safety and the presence of the police and other security facilities. This can be seen in an example from Fine et al. (2003), who found that urban youth, and especially young men of colour, express a strong sense of betrayal by adults and report feeling mistrusted by them. Given this, while for some groups, perhaps of older people, police on the street can seem reassuring; for others police patrols as well as CCTV cameras can be a constant reminder that such things are needed. Furthermore, and this is the thrust of Fine's et al.'s study, when trust in police is low, they can be seen as a threat in themselves. This will be particularly so for those who feel they may be victimized by the police, and this will typically be the poorer and minoritized members of the city. For example, Valentina in Rio says, 'Police officers are violent to those from a poor background.' During 2013, when the 'Black Lives Matter' campaign started, and later, in particular during the protests against police brutality in 2020 and 2021, it was clear that in many American cities many African Americans perceive the police as a threat.

Though race is a critically important factor in differential treatment by the police, factors other than race can also play a part. How safe and secure one feels in the city varies considerably from person and person and often relates to their perception and experience of their fellow city dwellers' behaviour in public space. In Rio de Janeiro Neusa, 59, whom, as we have mentioned, we found begging, also makes a modest income by collecting aluminium cans from the garbage throughout the city and selling them for recycling. She reports that she and her grandchildren are not safe, for there is always violence around.

In our previous book *Disadvantage* (Wolff and de-Shalit 2007) we discuss how risk to one functioning can become corrosive, meaning that the risk can spread to other functionings. For example, it is often the case that people on low incomes stave off the feeling of hunger by eating very cheap, highly processed food, including low quality pasta, rice, and bread, which is rich in sugars and carbohydrate. In overcoming their hunger, they risk their health, showing how achieving one functioning can put others at risk. But we also discussed the more obvious type of case where failing to achieve one functioning can also put other functionings at risk. Neusa has not been able to achieve secure housing in a safe area and thereby is more likely to be attacked in the street. In the terms of the capability approach, this puts her bodily integrity at risk. The situation is problematic for anyone without a secure home but is particularly difficult for rough sleepers. Another interviewee in Rio, Quésia, also mentioned above, is a rough sleeper and says she is terrified of paedophiles stealing her daughter. Because they sleep in the street, both she and her daughter live in perpetual fear and high risk.

The experiences of Neusa and Quésia, as noted above, contrast sharply with those of Mario, a Rio professor, even though they live in the same city. He reported that he loves strolling in the city and describes what he does as 'walk *freely* around the city', though he was acutely aware that others do not have his privilege of security.

The fact that people who live in different neighbourhoods experience the city differently intersects also with life stage, and the differences may be more intense for those in teenage years, looking for sources of entertainment and stimulation. One constant refrain in the interviews was the availability of facilities for young people. Parents in under-served areas of the city worry that their children may have no option but to hang around on the streets getting into trouble. Wealthier parts of town may have more facilities, such as clubs, leisure centres, or cinemas, which will often be another form of spatial segregation. Equally, many adults enjoy strolling around lively, historic, or beautiful parts of their own city, seeing street performers, or interesting graffiti, or going to the park and enjoying nature or solitude. Yet these opportunities to be amused or astonished in these ways may not be available to all in some cities.

It is no surprise that issues around safety and security were a constant refrain in the interviews, both explicitly and implicitly. Here are some further examples:

Mayowa (19) in Oxford says:

> There's always somebody somewhere, you're never alone on the street and there are lots of coffee places and the like. I haven't entered all of the coffee shops yet, of course, but I just like the atmosphere. It creates a nice feeling when you're

> just walking around and there are people everywhere talking, drinking coffee. I like that.

Chino is 65. In the 1990s he moved from Suriname to Rotterdam. He complains:

> It's crazy, I'm a grown man and I'm scared of walking down the street in certain areas. I would never return to Suriname, I guess whatever happens here is so much better than there, but it's really hard these last few years.

Omri is a 28-year-old student who has been living in Jerusalem for four years. He believes security should be the focus of the city's attitudes to its city-zens:

> The most important thing is the sense of security. I too, as a man, am not enjoying walking around in my neighbourhood after dark. I'm not even thinking about the old city [where religious and national tensions are high and often lead to violence], where I used to hang out a lot.

Sara is a 20-year-old religious woman in the neighbourhood of Pat, perhaps the least affluent area in Jerusalem, with a high poverty rate. She used to live in an exclusively religious neighbourhood, but moved to Pat because she enjoys the variety of people around her. But the downside, she testifies, is that she never walks outside after dark, whereas in other neighbourhoods, she says, women can walk outside, even alone.

3.5.3. Identity, Community, and Anonymity

Urban identity has a type of inbuilt duality or hybridity, which for many is one of the main attractions of the city. It can offer a sense of secure community, yet it also offers something diametrically opposed, as we saw that Simmel emphasized: the opportunity for anonymity, which itself has upsides and downsides. This multiplicity is beautifully expressed by David (33) from Berlin. He first says how much he appreciates that the city allows its people to be themselves and how there is no pressure to conform:

> Berlin is a mixture of many smaller cities and cultures. And with culture I don't only mean 'foreign' cultures but Germany, all corners of the country. (. . .) there is no 'standard model' here. You can develop and follow your own art of living . . . be authentic. It's all about diversity and differences. I like the openness and the curiosity that is created by this plurality of cultures and lifestyles.

However, he is quick to add:

> Variety has its downside too. Arriving in Berlin can be tough because everything is so impersonal here.

As for anonymity, it is interesting that our interviewees distinguished between not really knowing anybody, and respecting privacy. Moris (26), who currently lives in Amsterdam, but came there from a small town in East Holland, describes this as a function of diversity, and sees anonymity as an important condition for facilitating personal change:

> What I like the most about Amsterdam is the diversity. (. . .) Where I come from everyone in the streets already know something about you, so it's tremendously hard to reinvent yourself.

And Carl, 68, from Amsterdam says you can feel anonymity even if you do know many people:

> I enjoy the anonymity here. And it's not because I don't know people here, but because privacy is respected here. No one is actively at your back, no one looks through the window at what you do or where you're at.

Andrea (47) is a Berliner:

> I like the diversity, that everyone can be as they are. You can be yourself. (. . .) It's a general atmosphere I'd say, hard to describe. But, for example, you can be dressed however you like and people here meet each other more or less on eye level, I'd say.

But at the same time, another interviewee in Amsterdam, Maaike, 60, praises the personal feeling of knowing people around you: 'I adore Amsterdam. (. . .) it really feels personal—I know everybody and everybody knows me.' And Mano, a pensioner in Jerusalem, answers our question whether Jerusalem is a good city to grow old: 'I've got all I need. My friends are here, my family. Everybody knows me here and I know everybody, except for the very young guys.' And Carl from Amsterdam, whom we quoted above about the possibility of anonymity, adds: 'People here are willing to help in need.'

So there are two ways to experience the city: as an escape, as a place that worships privacy; but also as a place which has a great potential for a sense of community, bonding, and friendship. And these two ways can be experienced simultaneously; they are not either-or; many city dwellers feel that what they like about the city is that at any given minute you can choose how to experience it with regard to community versus anonymity. If so, a city of equals should be able to offer both ways to experience the city to all of its city-zens, with the emphasis on 'all'. An example of how unequal this can get is described by an interviewee who now lives with his wife and children in Berlin but originally immigrated from Peru. He describes a different experience, in which an immigrant is clearly recognized as an immigrant, as an outsider, under pressure to conform:

> When I go around, I speak to my kids in Spanish, to my wife in English—but I don't feel comfortable speaking that in public. I had an experience with racist attitudes. A woman told me to speak German in Germany, in her country, a woman in her mid-50s.

He nevertheless claims that there is no xenophobia, yet this experience has made him 'feel uncomfortable speaking other languages in public. (. . .) my wife's family name is German, so my kids took on her name, to avoid discrimination.'

Presumably nearly all egalitarians will sympathize and suggest that until everyone feels comfortable to speak their own language freely in public the city is not fully egalitarian. Another, potentially more controversial, example comes from Jerusalem where there is a large ultra-Orthodox Jewish community. Many of the men do not work but study the Torah until they are in their forties and then rely on welfare benefits. This creates some tension with the secular population, who struggle to work and who pay most of the local tax and national income taxes. In addition, some ultra-Orthodox city-zens argue that for them to be able to practice their beliefs and culture and to really feel a sense of community they need the city to respect their preferences about how others should behave. They want the public space to reflect Jewish values and since they believe there should be no transportation on Shabbat (Saturday), they want the city to forbid any kind of transportation on Shabbat, if not in the entire city then at least in the neighbourhoods where they reside. The religious prohibition on using vehicles on the Sabbath already causes tension with those who need to cross the neighbourhood using their cars. But much more controversial is the ultra-Orthodox community's demand that in some of the streets where they live the city should enforce strict modesty regulations including separation of men and women in public, including walking separately on the two sidewalks, one for men one for women. Feminists and egalitarians have argued that such regulations should be illegal as they discriminate against women. The reason members of the ultra-Orthodox community want women to walk on the other side of the street is that women are seen as sexually tempting, and men should avoid being tempted. This, feminists and egalitarians have argued, reduces the woman to her body and projects on women what goes on in men's minds. Ultra-Orthodox Jews answered that this argument restricts their right to culture and to a sense of community in their city (Ben-Dahan 2017). This is a just one example of how respecting everyone's sense of community, as well as cultural rights and groups rights, is fraught with difficulties. In Jerusalem restricting the use of public space by women is illegal, and yet in some of Jerusalem's ultra-Orthodox neighbourhoods it is the norm in practice, even if it is legally forbidden, and the local police look away.

Contested cases, such as the one just discussed, show how hard it may be to make everyone feel completely at home in the city, and there may be a need for negotiated compromise and accommodation so that everyone can get at least part of what they need to attain a secure sense of belonging. For many of our interviewees being able to attain both a feeling of anonymity and community was critical to their idea of a city of equals, and the ability to choose between anonymity and community and be whatever and whoever you want to be is perceived as freedom and autonomy. Amar moved to Amsterdam from Tunis thirty-eight years ago. He is now 65. He says:

> What is good here is that everyone is free. You can choose to do and to be whatever you like. People are very kind here, always helping each other. There is an atmosphere in the street that allows you to just talk or to feel free to ask for help.

Sivan, a 35-year-old dog walker who moved to Tel-Aviv from a Kibbutz (a small community in which many means of production are owned collectively) seven years before our interview and mentioned that one of the reasons she didn't get along there was that she was not comfortable with the fact that everybody in the Kibbutz knew everything about her. She likes Tel Aviv because it offers her privacy, though she does like the fact that people around know her:

> It's a matter of degree. People do not go on looking for you here, for what you're doing, for what your Mom used to do. They can't ask my Mom where am I, and with whom I've brushed my teeth. It's a matter of degree. Here I can sit with some friends, sit with people in the dog-court, have a nice conversation or sit for a drink with my neighbours. But at the end of the day, no one is looking through my window.

Being oneself for many people obviously includes identifying with a particular culture, gender, and ethnicity, and many cities have adopted welcoming pluralistic attitudes and policies. Although the pattern is changing, cities have been slow to recognize that many people live outside traditional family structures, perhaps living in a same-sex relationship, or are transsexual, or living alone either by choice or fate. For example, many cities give housing priority to families with young children. While this is understandable, we also have to ask what message such a policy sends to those who will never be in that position. We also need to consider the relation between social or economic class and a secure sense of belonging. In many cities a working-class pride is developed, which helps those with lower salaries still feel they belong to the city's story. Referring to Liverpool, Tid (30), now living in London, explains why the working class who still work have this advantage whereas the unemployed, he claims, lose this sense of belonging:

> You are working class, you get payment. But you take care of your family. You'd work in a factory like Vauxhall [which is located in Ellesmere Port, near Liverpool]. And so you went and had that working-class experience with others. You're not doing a very skilled job but you brought home money and looked after your family. But if you never make it, if you're never part of this community then you'd go and get money from the government. And you were 'on the dole' as we used to say, you know, so in a sense you lose the working-class hero status. You're also thought of as lazy or drinking or someone who takes drugs—a loser really. (. . .) You lose that privilege to look down on the upper classes, you know, as a working-class hero, you lose the ability to say that the government doesn't give a fuck about you. You lose the working-class hero privilege once you're on the dole. So you are not a part of that community.

For many interviewees, a sense of community meant feeling oneself to be at the heart of the city, whereas lack of it implied inequality. Moti (61) has been living in Tel Aviv for over forty years. He expressed this very eloquently:

> Let me explain it to you, that alienation here, this is the most alienated place there is. Every neighbour to himself. If he does not die and stink you will never know that he is dead. It is not like it used to be, that people would knock on your door, say hello. This is not a cliché. People would sit at my home, laughing. You would see shining eyes (. . .) Now we are like wolves to each other.

'But isn't it what is so good about the city', we ask him, 'that people can avoid others?'

'Yes,' he says, 'but Tel Aviv took it to the extreme. (. . .) Consider Jerusalem, there is solidarity, there is care. Not only among religious people for religious reasons.'

4

Interview Themes and Results, Part 2

In this chapter we continue our report of the results of our interviews, referring to the two remaining core values which construct the idea of a city of equals: diversity and social mixing; and non-deferential inclusion. We review suggestions by our interviewees that relate to the notion that a city of equals requires consideration for people in their differences rather than offering a 'standard package' of resources and facilities for all. Thus we start with suggestions that special arrangements need to be made for different groups in the city.

4.1. Themes That Relate to the Value of Diversity and Social Mixing

4.1.1. Special Arrangements for Elderly People

In 2012 Ole Kassow, a city-zen of Copenhagen and a devoted bike rider, saw an old person sitting on the bench. He noticed that the person was sitting there the next day too, the day after, and so on. A thought crossed his mind: that while bicycles are the most popular means of transportation in Copenhagen, elderly people tend not to use them. Out of empathy, he instinctively took a rickshaw bike and went into a nursing home and offered a ride to one of the elderly women there. What started as a spontaneous ride full of fun developed into an extensive global project of volunteers offering rickshaw bike tours to elderly people in various cities. (See Cycling Without Age 2023). It's about enabling elderly people 'to feel the wind in their hair again', as Kassow puts it.

Age inequality in the city is an important issue, not only in respect to the importance of ensuring a mix of available amenities, suitable for people at different life stages, but, critically, in terms of transport, especially at peak times. We've already mentioned that in some cities older people prefer busses to the underground, sacrificing speed for comfort and a more pleasant view. Some cities make special arrangements for their older city-zens. For example, some of Japan's cities have designed their sidewalks to allow elderly people

City of Equals. Jonathan Wolff and Avner de-Shalit, Oxford University Press.
DOI: 10.1093/oso/9780198894735.003.0004

to walk more easily with their mobility aids. Athens' sidewalks have a special path, yellow grooves, for blind people who navigate with sticks (Graham 2015; Greek Boston n.d.).

Heidi, 65, from Hamburg very much appreciates a bus shuttle arrangement there:

> Going back to the question of old people without cars, and without the opportunity to do their own shopping . . . I really have to say, we have a superb community . . . There's this bus shuttle that picks people up wherever they live and takes them to the market, where they can do their shopping, if they like they can do it together, and then they're driven back home. That's fantastic for the older ones who aren't that mobile anymore on their feet, nor by car.

Focusing again on relational equality, the elderly people we interviewed on the whole mentioned special arrangements less than the attitudes of others they meet in the street. Erika, 75 from Hamburg, said she experienced a lot of disrespect:

> That is really one of the worst things about this city these days. I cycle and there are people just standing or walking on the cycle path, and then I use my bike bell and the people just act like 'who is this old woman that she rings her bell at us', and I hear them calling me stupid words.

Petra (60) from Berlin said she found it more difficult, now that she was not young any more, to 'find activities and social contacts' in Berlin, and so she moved to a suburb north-east of the city where she can meet people she knows more frequently. Sabine (49) is still younger, but she agrees that elderly people, especially pensioners, are the most disadvantaged in Berlin, particularly those from the former East Berlin.

Tina (74) has lived in Oxford for twenty years. When asked why she likes Oxford so much she answers it's the fact that people of different ages can live together, which makes her feel comfortable:

> Well, it's the mixture of ages in Oxford, there's students, there's families, there's old people. And it's very lively. And I've been living in another town for a few years, and just came back here . . . it's much more lively in Oxford.

Carl, 68, is a pensioner in Amsterdam. He believes that older people should be able to enjoy the city's pleasures, either by special arrangements or because their pension is sufficient.

> Personally, I think that there are huge inequalities among the elderly here. . . . People who can't afford this coffee are missing out on all the fun! What's great about this city is the chance to have a good coffee and look at the rain through the window, to read the newspaper, to talk to strangers like we are doing now.

But Carl adds an interesting observation:

> There are two kinds of people in this world, the creative ones, and the passive ones. The city wants to encourage the creative ones, the entrepreneurs, because that is what brings in the money, and what makes a city special, with character. But it is much harder to save money when you are self-employed or when you are an artist. And so as the same people the city once encouraged grow old, they can't afford to live in the city they help create. I see some of my friends struggling, and it's very frustrating.

Obviously, a city which respects its elderly should not make it harder for them to continue enjoying the city's pleasures.

4.1.2. Special Arrangements in Respect of Young Children and their Parents

When we ask Sulaika (female, 36) who lives in Acton, London, to name three issues that bother her with regard to inequality in the city, she points to homeless people, the high unemployment rate, and that 'children's centres have been shut down, so families that would have been supported through these facilities are now facing problems'. Indeed, special arrangements are also needed at the other end of the age scale, and especially, for very young children in prams and pushchairs, which implies, of course, that the special arrangements are for parents with young children as well; primarily, though not exclusively, younger parents. Making provision for younger parents can be important in terms of equality. Young parents often find it difficult to cope: with making ends meet, with the change in their lifestyle when they become parents, especially coping with far less sleep than they may have become accustomed to, with the responsibility of taking care of infants, and, for many, with the challenge of building their own relationships with one another while meeting the various other challenges of parenthood, at work, and so on. But life is likely to be even more difficult for single-parent families, and in such situations the parent—almost always the mother—is likely to be faced with financial pressures as well. The city can either ignore, or make minimal extra provisions, for young parents, whether in couples or alone, or make a special effort to ensure that young parents feel that the city has understood and embraced their needs. A city that does embrace young parents' needs indicates that it understands that a thriving city should take care of the reproduction of its own population and make itself welcoming and more comfortable for families with young children, rich or poor.

What can a city of equals do in that respect? In an illuminating interview Erika (75, Hamburg) said that had she been a mayor for a day she would provide not only free day care for working parents, but also night care, for parents who work night shifts or who wish to have an evening off to meet friends, go to the pub, or whatever.[1] The city can also offer subsidized places in kindergartens and by designing and building new playgrounds, free libraries, working with, for example, churches, to allow the parents and toddlers to use the place in the mornings for meetings, especially on rainy days, and so on. But from the interviews we learnt that the city can do much more, in paying attention to small details. Here we have in mind designing features such as the width of corridors and doorways in public spaces, as well as lifts, and safe places to cross the street, as well as, once more, adequate supply of well-maintained public toilets. Sometimes the difficulty of navigating a pushchair can lead to considerable delays and frustration, and often a feeling that one does not belong in a particular place. Klaus from Hamburg says:

> Fifteen years ago, I had to take care of a child in a pushchair. And I couldn't have imagined how badly the city was equipped to accommodate for people using pushchairs . . . not even the central station. The city was not at all accessible. But now this is all much better, now they have provided accessibility everywhere. (. . .) I visit Berlin quite often and I am shocked (. . .) the sidewalks are in a terrible condition, that's something we don't have here in Hamburg at all as a problem.

Nicole (36), a Berliner, argued that not having enough activities for children is a clear sign of inequality—not only between children and adults but also between childless families and people with children.

Sabine, 49, also from Berlin told us how she had to move outside of the city because she felt her child was not secure in the city. She thought that in Berlin single parents are disadvantaged, as well as children:

> I was a single mum, back then I moved to the almost-countryside in order to raise my child. Here in Berlin, I think this would be much harder. I think it's more

[1] This, in fact, is similar to proposals presented by the British feminist movement in the 1970s, with regard to their demand for twenty-four-hour nurseries. See Phillips (1985).

> dangerous here, there are more risks and threats. So, you can offer fewer freedoms to your child.

On the other hand, Seyneb (34) from Hamburg, believes that one of the parameters which make her city egalitarian is that

> Hamburg is very child and family friendly. There are many offers for children, especially in Finkenwerder [a small quarter in Hamburg] . . . and child care is for free.

When thinking of what makes (part of) a city child friendly we include urban design, public policies, and social norms, allowing children to move through the city and play. Critical factors include safety,[2] not only in terms of reducing the dangers of traffic, but also architectural proportions, such as making sure that street furniture is usable by children in non-hazardous ways (Brando and Pitasse-Fragoso 2022). A city of equals invests in services for all its children and, especially, those who are more disadvantaged and lack access to private facilities.

Now, when it comes to education, there are many children with various needs and preferences, although of course it is the preferences of their parents that typically carry weight in these areas. Therefore, providing a varied and flexible range of educational opportunities for children with different social and academic needs, and from families with different preferences, seems a key feature of a city of equals. Jenny, 28, praises Tel Aviv:

> Here there are many possibilities, many types of schools, alternative schools as well. There are all sorts of schools, like a democratic school or an anthroposophical school that you won't find in other cities.

However, Ma'ayan (40) remarks, cynically, that when it comes to kindergarten and the first classes, the city fails to provide adequate education for 'children whose preference is not to engage in bullying'. Recently, the city of Tel Aviv found itself in a crisis. Because rent prices rose rapidly, by more than 10 per cent annually, and because the salaries of teachers are typically low, many teachers left Tel Aviv and moved to other cities where rents are much lower. Some schools, especially those with constrained budgets, have to employ less-experienced or less-qualified teachers; to the extent, that Ram

[2] Hood (2004) reviewed methods of evaluating children's rights in a city. It is argued that there is a correlation between children's well-being and their safe access to public space. Fear of traffic and lack of access to appropriate play spaces might have adverse impacts on children's general health and emotional well-being.

Cohen, headmaster of a well-known high school in the city declared that 'we simply betray our children' (Shany 2022). One natural response to a crisis in state education is to encourage the growth of fee-paying private schools, but of course a division between state and expensive private education is a typical mark of a city (or indeed country) of inequality. The private school system in the United Kingdom, for example, is very often criticized for cementing class division and reproducing privilege.

One common pattern in many cities, owing to high housing costs and limited space—both inside and outside space—in constrained apartments, is for families to leave the city when they have their first child. Under such circumstances the city will have many younger residents, in their twenties, and many retired people, or those still working but whose children have left home, in their fifties and sixties, but a smaller number of families with young children, especially those on a medium income. Such is the situation in Tel Aviv, for example, one of the most expensive cities in the world, or in New York City. In New York a young woman told us she had a perfect job, working for the *New York Times* and earning a good enough salary. However, her plan was to continue working in New York until she marries and has a child. Then, she said, she would leave the city—her plan was to move to Princeton, New Jersey, where schools and raising a child are more affordable and where renting was cheaper than in NYC. In order to keep families with young children in the city, some cities have introduced rent control, or even made buildings available for public rent, especially for young parents, but many cities need to do much more in this respect.

4.1.3. Women-friendly City Design and Planning, and Gender Equality

Raising related concerns to attitudes to the elderly and the young is gender equality, by which we mean here equality between men and women, or, more precisely, men and women as socially constructed in the city. We discuss other aspects of gender and inclusion of LGBTQ+ communities and individuals later. In the literature, equality between men and women in general, not necessarily in the city, often focuses on income discrimination in salaries, and in opportunities for jobs or for administrative positions, which without doubt are of critical importance especially at the level of national law and regulation. But in the city, creating more equality between women and men is a good example of how cities can cater for equality without transferring money from one person to the other, but rather by creating the right atmosphere, investing in infrastructure, and also by paying attention to details, such as by asking questions such as why don't women make use of the public park around the corner?

Naturally, because this theme, gender equality, importantly includes how women feel, extra attention should be given to how women experience the city: to subjective measures. But objective measures of gender equality in the city matter too, such as the participation rate of women in the labour force, and especially in better-paid professional and executive positions, or whether women run for offices in the city. In recent years there has been a rapid growth of women who have been elected as mayors, in Tokyo (Yoriku Koike), Sydney (Clover Moore), Surat, India (Ashmita Shiroya), Bucharest (Gabriela Firea), Madrid (Manuela Carmena), Barcelona (Ada Colau), Paris (Anne Hidalgo), Amsterdam (Femke Halsema), Copenhagen (Sophie Hæstorp Andersen), Oslo (Marianne Borgen), Stockholm (Karin Wanngard), Rome (Virginia Raggi), Berlin (Franziska Giffey), Zurich (Corine Mauch) Geneva (Frédérique Perler), Sofia (Yordanka Fandakova), eThekwini, SA, (Zandile Gumede), Montreal (Valerie Plante), Chicago (Lorie Lightfoot), San Francisco (London Breed), Seattle (Jenny Durcan), Washington, DC (Muriel Bowser), Boston (Kim Janey), Atlanta (Keisha Lance Bottoms), and many more cities, though we are still very far from numerical equality. Steps to go further are being taken in various ways. For example, in 2012 a national law was passed in Nicaragua, according to which parties were subjected to gender parity in the submission of the candidates' list for municipal elections. This had impressive results. If in 2008 only 8.6 per cent of the mayors were women, following the implementation of the 2012 law, the figure had risen to 40.1 per cent (Gender Equality Observatory n.d.; National Democratic Institute n.d.). Similar regulations adopted globally could be part of a process to improve equality between men and women in political positions in the city, although this movement is still in its early stages.

Returning to subjective measures, women and men often report their experience of the city in general, and public spaces in particular, in different ways. To examine this, we conducted an experiment. Students in Jerusalem were asked to walk in couples, a man and a woman, in various areas of the city, in different times of the day and night. For example, they observed the main food market very early in the morning, at 5 a.m., and the bars and pubs area around midnight, in several different parts of the city. They were then asked to report in class about their experiences and how they felt during these walks. The reports were markedly different. The male and female student watched the same urban activities together—trucks bringing fresh fruits to the market, drunkards leaving the pub, parents with their children in the playground, people shopping in the mall—and the same people—homeless people, merchants, undocumented (migrant) workers, groups of friends socializing in pubs, passengers rushing through the main train station, and so on—and yet their subjective experiences differed radically. Often male students described the experience as a series of adventures, as fascinating and eye opening, pointing out moments of beauty (sunrise over the roofs, a truck full of colourful apples arriving to the market, a merchant movingly singing a traditional prayer, fathers and mothers pushing the swings in the playground, and so on) whereas the female students often described a sense of anxiety, fear, or uneasiness when unexpected or boisterous events took place very early in the morning or very late at night, or often pointed to the misery they saw—the undocumented worker hiding from the police patrol, people shouting at each other, or the experience of being the only woman in an area.

This experiment with the students suggests differences in the way women experience the city and how they perceive it; but some might contend that it does not necessarily imply inequality. And yet, connection with equality seems obvious and compelling. Out of concern for their safety and peace of mind, women often have to restrict their freedom to stroll after dark because even though laws in theory protect them, the local social norms that should accompany such laws are too thin to be relied on. Iniko and Kiki are two young women in Rotterdam. Kiki says:

> It's not like we can walk to whatever neighbourhood we want. I try not to walk alone at night, and when I do, I'm always on the phone or pretending to be on the phone.

Iniko adds:

> I have all kind of tricks. To pretend you're on the phone, to be ready to dial the police. (. . .) There is a really big difference between Rotterdam by day or by night. I would tell you to wait and see, but don't. (. . .) They'll follow you, shout at you, chase you.

The inequality described by Iniko and Kiki is that, unlike men, or at least to a far higher degree, they experience fear and are subjected to harassment. Maaike (60) is from Amsterdam. She has been living in the city for forty years and enthuses about it:

> I adore Amsterdam. I love everything about Amsterdam. I'm an artist, so most of all I love how aesthetic and beautiful the city is, and that means a lot to me. I like the culture and the biking.

She seems like an open minded and egalitarian person:

> I really enjoy the way immigrants changed Amsterdam. There are now so many interesting food shops to try, such unique art and new fashion. I love it.

But when she thinks about inequality between the various neighbourhoods in the city, she says that there are neighbourhoods where 'it will be dangerous to bike alone after dark, as a woman'. She tells us that 'I probably wouldn't bike alone in a big park out of the central area.'

Rivka (47) from Jerusalem is a Jewish religious woman who was born and raised in the city and has only spent a few years of her life elsewhere. She claims that the experience of fear and uneasiness in the city varies according to age. Now, she claims, she has reached the age when she feels more secure, and, on the contrary, is looking for opportunities to meet people who are different to her, from other ethnic and social groups. She says:

> The city is the 'real thing', it exposes me to different people. Heterogeneity is positive, it enables you to get out of your bubble. Here is an example: I go to the clinic and the entire staff is Arabic.[3]

But, she adds, it all depends on your age and status. Accessibility is important, but much more important is feeling secure and safe wherever you go in the city. Younger women, she emphasizes, do not feel safe as she does. They suffer from harassment. Shira (25), from Jerusalem, who has just been to London for six months, reflects about the two cities. 'What do we mean when we say that a neighbourhood is not a good one?' Her reply is interesting. She claims that women's feeling of insecurity in a neighbourhood is correlated with their perception of the streets as dirty and not tended with care.

Equality between men and women in the city goes beyond a sense of gender security, however important that is. Vienna is often pointed out as a pioneering city in 'mainstreaming' gender equality. One evocative example is the response by the Vienna's authorities to a problem that must be replicated in cities throughout the world. Typically, women between twenty-five and forty years of age find themselves with primary responsibility for driving their children to the kindergarten and school. Alas, heavy and unpredictable traffic makes it very hard to ensure a regular, prompt, arrival at work. This has many consequences, including, it is thought, delayed promotion at work

[3] The medical professions are very popular among Israeli Arabs. According to the Ministry of Health, in 2021 46 per cent of the new registered doctors were Arabs (who are 21 per cent of the general population in the country (Doctors Only 2021)). We do not have figures about Jerusalem, but the figures are probably higher.

because bosses tend to regard women as less reliable, on the basis that occasionally women fail to arrive at work on time. Now, there can be numerous ways of trying to tackle this, for example, by trying to educate men, and their employers, so this burden is shared between spouses. But Vienna's approach is very interesting and shows how municipalities can think out of the box and help to mitigate such issues pending broader social change. Their idea is to place more kindergartens close to areas occupied by families with young children, so that parents, and especially mothers, can walk their children there and not get stuck in the traffic, or to build educational campuses near centres of employment, so that parents can combine the journey to work and to school, and also can visit their children often during work (Damyanovic 2013, 60, 92). They call this approach 'a city of short distances' (Damyanovic 2013, 25), which is not only about reducing travel, but also about changing the life tempo, slowing down the way we live in the city, not having to rush from place to place. It is very close to the idea of the 'fifteen-minute city' which is now being widely discussed in urban planning. The idea of slowing things down was reflected in one of the interviews. Linn (25) is a Londoner who works in the film industry. When asked about inequality in the city she mentions both race and gender. What would you do to change this? Her answer is interesting: it is about fostering a sense of community by slowing down things, including, she claims, the very fast rate in which people now move home from flat to flat, which means communities never really form:

> So, yeah, in this sense . . . what I would like to see more is, you know, more community feel, and have people, you know, sharing cars and even gardening tools and having more incentives to have like a community. But then again, that would be quite difficult because people are moving so quickly around in London. So, it's difficult to establish this kind of community.

Tamir (40) from Tel Aviv confirms this sense of slowing down as an advantage which should be shared by everybody. 'Since arriving to Tel Aviv, I have been using my feet again, I walk everywhere'. He describes this as a liberating experience.

Another example from Vienna was to put volleyball nets in public gardens which before only men had used, playing football. Once the volleyball nets were installed women started to play there too (Förster et al. 2021, 38). Indeed, some of the female interviewees mentioned equal use of parks for sport and recreation as a matter of gender justice.

Another obvious measure of 'gender mainstreaming' is gender-neutral road signage, in which icons representing an abstract person are not by default male, but, for example, show a woman, or a woman and a child (Wander Women Project, n.d.). Such strategies are often used now by urban planners, to actively counteract gender bias in policies and regulations, and to promote more equitable relations between women and men (Verloo 1999; 2000, 13). But of course, as we have already seen, the issue is not only road signs, but also street, public space, and traffic planning, so that it appeals to, and is comfortable for, the use of both men and women. Planners in Vienna measured the average speed of walking adults and the time needed to travel 1 km (9–13 minutes) and compared it with the time when accompanied by toddlers and children (24 minutes) (City of Vienna n.d.), as well the time for people with severe mobility problems (McManus 2020). They then thought how such information should influence and modify planning public space, for example, where benches should be located for people to rest. Interestingly, in the TED talk we referred to in Chapter 1, Janette Sadik-Kahn, transportation commissioner of NYC between 2007 and 2013 tells how when she entered the job she noticed many people perching on fire hydrants because NYC was a 'city without seats' (Sadik-Khan n.d., minute 2:58), which inspired her to spread more benches and even hundreds of lawn chairs around the city (5:30).

Returning to Vienna, the city also found that women walk and use public transport more than men, whereas men use cars, motorcycles, and importantly, bicycles, more than women. For example, 24.8 per cent of men report walking every day in the city, whereas 31.5 per cent of women do, while 33.3 per cent of men report using cars or motorcycles every day whereas only 24.1 per cent of the women do. So the city concluded that a move away from cars to public transport and investing in sidewalks will have a positive gender equality effect. Of course, this is only one of many good reasons to encourage people to use their cars less, but not all initiatives will improve gender equality. For example, if men are encouraged to ride bikes rather than drive their cars, and sidewalk space is consequently used to create more cycle lanes, then Vienna's own research suggests that it could harm gender equality by taking space from women walking to give to men cycling. Which complicates things, but who says life is simple? The city of Vienna suggested five principles for gender mainstreaming (Förster et al. 2021): gender-sensitive language; gender-specific data collection and analysis; equal access to and use of services; equal participation of women and men; and integration of equality into steering instruments.

It is interesting that even though the data from Vienna, if generalizable, suggests that women walk in public space more than men, several female interviewees in different cities see public space as dominated by men. Kiki from Rotterdam expressed this straightforwardly, when she said: 'I sometimes feel that there are so many more men than women in this city'. And Annette (33) and Revital (36), two women from Tel Aviv, complain that the city does not invest in public toilets because:

> Men can pee anywhere, on a tree, for example, it's not illegal (...) even taxi drivers, stop and pee in the public garden and women, who tend not to do so, cannot find a place to urinate.

Michal (40) a woman from Tel Aviv, in response to a question about what is unequal in the city, replies that the authorities cut down trees and therefore there is not enough shade.

> You [meaning a woman, she uses the female verb in Hebrew] walk in the street, you need [again, she uses the female verb] shade, you need shade. And they trim the trees, they cut the trees. I know they have to trim them, but you walk in the street, you need shade.

So again, Michal refers to decisions and policies that are taken by the city authorities which unintentionally and perhaps unconsciously do not consider the perspective of women. It is, perhaps, unusual to see a gender difference for the need for shade in the street, but this is certainly the perspective that Michal takes.

The Vienna document concludes that the city's services and products can only be designed to meet everyone's needs in terms of the five principles of gender mainstreaming if the city has data on both women and men and how they use these services. But perhaps more is needed to ensure that the services are equally accessible to both men and women. Not only should planners double check whether the frequently different circumstances of women and men and the different living conditions of women and men are considered in planning and designing services, they should also try to involve women and men equally in committees and decision-making. And yet, having said all that, we need to keep in mind that gender equality is only one component of a city of equals. Lisa, (43) says that she is worried how she will survive economically speaking. Reflecting about her city, Oxford, she argues that we must not think that gender issues make up the full picture:

> I definitely think that many people here, young people, are very aware of the gender and sexuality issue—but it seems to me that there are many other problematic things that they are not so much aware of. It's very difficult to formulate this. I think

> Oxford as a city and as a university is trying to be very inclusive . . . and promoting equality . . . and people are aware, but there is awareness of certain types or directions in particular. Social and especially financial inequalities, I think that's what I am mostly concerned with.

This is a useful corrective for our emphasis on relational equality; it has long been a concern that the rhetoric of relational equality—'we are all equals here'—can be used to screen highly damaging material inequalities. While we do not think attention to financial inequality diminishes the importance of relational equality, it reminds us that material factors cannot be swept aside. However, material inequality will typically exert its effects through people's differential life experience.

4.2. Themes that Relate to the Value of Non-deferential Inclusion

To recall, our somewhat unusual notion of 'non-deferential inclusion' is to be granted access to the facilities and privileges of a city as a matter of right or entitlement, on the same terms as others. Failures of non-deferential inclusion can come in many forms. At its most crude, it can be a simple matter of exclusion, such as lack or denial of rights. A more subtle failure of non-deferential inclusion, and what we will focus on more here, is what can be called 'deferential inclusion', where access to facilities and privileges of the city are granted, but on more onerous terms than for other city-zens. This could mean having to throw oneself on the mercy or discretion of officials or other gatekeepers, or always having to wait longer than others, or to go through bureaucratic hurdles, or being made to feel that others are somehow doing you a special favour in giving you what you are entitled to. To suffer deferential inclusion is to be made to feel a second-class city-zen, even though you do eventually and with difficulty receive everything to which you are entitled. In what follows we will point to features of the interviews that show how people felt that the city can sometimes fail to achieve non-deferential inclusion for all, but will reserve fuller analysis for Chapter 5.

4.2.1. Communication beyond Transportation: Words and Vision

How often have you found yourself looking for Wi-Fi in public spaces, and feeling the frustration of a service that claims to be functioning but doesn't actually allow you to send an email or reach a website? Obviously, access to the Internet while on the move in today's world is a must, and for many people who cannot afford a comprehensive data package Wi-Fi has become essential. In many cities where cash and notes are not used any more, one might find it very difficult to pay one's bills or pay in shops that lack access to the Internet, for which one often needs access to a good Wi-Fi service. It is a most basic need in contemporary economy and society. This was clearly noted during the Covid19 pandemic, when people relied on Zoom and similar communication systems in order to communicate with their relatives and friends, and on Internet services to buy food and other supplies, and those without a laptop or smartphone were especially isolated. But even in everyday life we have become dependent on access to the Internet and Wi-Fi services. Some pioneering cities in that respect, including Buenos Aires, Boston, Baltimore, Montreal, Quebec City, Taipei, Beijing, Kuala Lumpur, Vienna, Helsinki, Malmo, and Geneva, have introduced free and reliable municipal wireless networks across the city. But in other cities the services are much more variable, often offering free service in airports and, sometimes, downtown and to those living in more prosperous neighbourhoods whereas those living in poor neighbourhoods, not to mention semi-legal buildings, favelas, and the like, lack such free services.

In Chapter 3 we paid a great deal of attention to transport. But communication in the city should be perceived as more than just the ability to move from A to B. Rather, communication should be perceived as the basis for any sense of place, of belonging and community in the city. Communication with others around us is about bonding, about grabbing the opportunities that cities offer for mingling, socializing, making friends, having a sense of community, and, through all these, giving meaning to what we see and sense. Urban communication is therefore both exchange—of information, ideas, views, opinions—and the infrastructure which enables such exchange. Urban communication is therefore a form of connecting people, through dialogue, which takes place in various forms: speech (among city-zens and between the city-zens and the authorities), advertisements, through the new media but also by hanging posters at the entrance to the local grocery or convenience store, and through public means of transportation: trains, trams, buses, bicycles, and the like. It also includes, and we will return to this, how people look at each other.

City dwellers we interviewed seemed very concerned about communication in the city, and expressed a sense of frustration and disappointment when it did not work as they wished or did not offer opportunities to engage with others. When Nicky (36) describes her city, Oxford, she complains:

> People always seem busy, running from one place to the other, never looking up and around them, they don't talk to each other or mind the people around them that much.

It is for this reason that the city of Copenhagen developed an attitude to communication which goes beyond transportation, and which we find appealing and egalitarian. They believe that equal access to communication creates a sense of community and equality between all city dwellers. It all begins, they argue, with *eye contact* (Saaby 2015; McLaren and Agyeman 2015, 137–8). So, for example, a regulation forbids new businesses and new public buildings from using black or mirror glass in public-facing walls above the height of 120 cm. The idea is to create eye contact between those inside the building and those walking by, and the regulation removes a possible (literal) barrier. When people sit (inside the cafe or the building) their eyes are around 120 cm above the sidewalk's surface, rising to 160 cm when they stand. The result, these planners argue, is that those walking by do not feel unwanted or unwelcome, and they also give greater respect to those sitting inside. (In the words of Tina Saaby, they do not pee on the building's walls.) As UN-Habitat's executive director Joan Clos writes in a preface to UN-Habitat's book *The City at Eye Level* (Karssenberg et al. 2016):

> UN-Habitat emphasizes the role of streets and public spaces as a connective matrix on which healthy and prosperous cities must grow, embracing the essential requirements of being inclusive, connected, safe, accessible, multi-functional, and liveable. Therefore, the quality of the ground-floor façades we pass close by at eye level is particularly important to enhance environmental sustainability, enrich the quality of life and promote equity and social inclusion. Tools and regulations to strengthen the relationship between the ground floor and the street will improve the interact.

Like others have done, most notably, perhaps, Philip Pettit in his work on Republicanism (Pettit 2014) which incorporates what he calls 'the eyeball test', we therefore adopt the concept of 'eye contact' to express a notion of equality, although for us it is a particularly important aspect of the egalitarian city, both literally and as a metaphor, for the proper relationships between city-zens. Being able to look each other in the eye is a precondition for people to feel that they are equally respected and that they equally belong to the public space of the city.

In 2016 we both taught a summer course in Dubrovnik. We asked our students to talk to people in the street, asking them what would characterize an egalitarian city, and return to the classroom and report their findings. Interestingly, one of the common answers was 'a city where we look at each other's eyes', meaning that we are sincere, honest, and open to others. This eye contact points to the intimate relationships between inclusion, feeling included and at home, and equality in the city. In a city where hierarchy rules as a social norm, those on the lower levels don't dare to look at the other's eyes, and those on the upper levels don't care to do so. Indeed, in our interviews sight and physical metaphors were often used in this context: Can you look others in the eye? Do others look down on you? Do you stand shoulder to shoulder with others? Moti (61) from Tel Aviv complains about how alienated the city has become and how this affects inequality. He says:

> We are all capitalist pigs, who don't care, we don't look people in their eyes, just run over them. (. . .) we lost solidarity; we lost the ability to watch each other and see each other.

This is a subtle issue of course, and context matters. One important domain of contact is when city-zens need to interact with those in a particular role, whether with a public servant in an interview for eligibility for a welfare benefit, or a private shop assistant when making a simple purchase. Here, typically, looking someone in the eye is a type of physical metaphor for honesty and respect, though we acknowledge that cultural practices differ, and looking someone in the eye, such as a police officer, can be taking as a sign of defiance or disrespect. Indeed, people often raised the issue of whether eye contact can have different meanings in different cultures, or within different groups in the same culture, when we presented our research at conferences or when teaching our seminars. For example in heavily religious cultures different practices can be expected, and, men might take care to avoid looking at women's eyes, as in Jerusalem.[4] In terms of chance interactions in the street, in some cities, such as London our observation is that unless people have children or dogs with them, giving an innocent reason for interest, few people will catch each other's eyes when they pass, whereas, for example, in Oxford, especially among the academic community, people are more likely to take deliberate notice of those around them, as the chances are much higher that people will know each other, and it is rude for a student not to acknowledge a professor and inconsiderate and arrogant for the professor not to acknowledge the student. But more than everything else, we adopt the idea of a city with eye contact as a metaphor for communication and honesty between city

[4] This is a different aspect of inequality. We thank Tal Eldar for this important comment about religious men in Jerusalem.

dwellers, just as 'standing shoulder to shoulder' is rarely called for in practice but expresses an idea of solidarity.

4.2.2. Inclusivity

The discussion of community has already introduced the next set of issues, namely how inclusive the city is, which we will now focus on more directly. Broadly, we could say that inclusivity was raised in two often-related contexts: one about minorities in general—religious, ethnic, sexual orientation, and so on—and the other concerning immigrants, especially in cities where immigration from other countries is particularly common. Some cities, most notably huge cities in China such as Beijing and Shanghai, are struggling to cope with massive numbers of migrants from the countryside or other smaller cities and attempt to restrict numbers by regulation, turning many millions of people into illegal residents. But in most cases, when locals talked about inclusivity and immigration, they had in mind people moving to the city from foreign countries, often not knowing the local language or customs, and bringing different values with them.

The arrival of immigrants can be a social challenge—they speak other languages, dress differently, and look different from those already in the city, or they have values and beliefs which differ from those already common in the city. Indeed, in several interviews people were not so keen on having immigrants in their cities; but such sentiments were raised less often than we expected. A cynic might say that people hid their real views in the interviews, and we cannot rule this out, but we believe that there is another explanation: that the discussion was usually set in the context of the city, and, as we argued above, people can change their attitude to immigrants when they switch from thinking like a state (or as citizens) to thinking like a city (as city-zens), and become much more tolerant, hospitable, and inclusive. This can even be true for people who express chauvinistic attitudes when they think of immigration on the state level, and are much more opposed to the 'abstract' immigrant than they are to the actual people from other countries whom they know and work with, sit next to at church or the football, or their children delight in playing with (de-Shalit 2018). We certainly would not claim that there are no voices against immigrants but in our interviews they were exceptional.

According to the Knight Foundation survey (n.d.) about what matters most in the city, the city's openness—how welcoming the community is to different types of people, including families with young children, minorities, and talented college graduates—came second only to social opportunities. This was reflected also in the interviews. Jason is 18 from Hamburg. He tells us this story:

> Do you know what happened on my first day at work? I work for a construction company. On my very first day, my colleagues met me and the first thing they ask me is 'You're not a Muslim, are you?' Can you believe that? What did they think I would do? Blow them up with a bomb? And now, well, now they know me . . . and they say 'Hey, Jason, cool guy!' but he's the exception. . . all other black people are still bad.

Inclusivity and exclusion can be a function of formal regulations, which can make it uncomfortable for some people to live in the city, as they have to choose between their own norms and values and obeying the city's regulations. It may seem strange now to think that there are regulations specifically designed to exclude or make people feel uncomfortable, but whether by design or accident some regulations can certainly have that effect. For example, Raze (24) compares London favourably with Geneva. To our question whether people in these cities meet each other on eye level, he replies:

> Yes, in London people with religious clothes, for example, can work in shops and in government—in their religious clothes. That's great, I think. In Switzerland that's not possible. Religious people can't wear anything like that for government jobs. So I think it's a strong sign of equality [in London].

This echoes a debate in French cities. In 2016, for example, Nice's municipality banned the use of the burkini swimsuit because—so it was alleged—it represents Islamic extremism, and yet a month later the court overturned the ban (Agence France-Presse 2016). More recently, in 2022, the issue was discussed in Grenoble, and the municipality authorized all swimwear, including burkinis. This, of course, is a consequence of the particularly French concept of 'laïcité' which calls for a sharp divide between religion and the state, including aspects of public life, and has often been interpreted as requiring a ban on religious symbols in public. The suspicion, however, is that it is implemented unevenly, and has been used as a tool to disproportionately target Islamic culture, while being much more tolerant of Christian and even Jewish styles of dress and presentation. In egalitarian terms it raises the fraught question of whether equality means 'sameness' or 'tolerating people in their differences'. It should be clear from what we have already said that our sympathies are with the latter, while recognizing that for both principled and pragmatic reasons there often have to be some limits to toleration. At the same time, we appreciate the value of shared civic experience but believe conditions should be put in place so it can develop its own forms organically, rather than be forced by law and regulation about such things as regulating personal appearance.

The example of dress codes concerns ways in which formal rules can exclude. However, as J. S. Mill taught us, social exclusion can also take place by informal means. Social pressure, even norms themselves, he argued, are no less exclusive than regulations. In fact, they are often less noticed and less concrete and therefore more difficult to oppose. Alba (38, male, who came to Berlin from Africa) tells us his story:

> I was refused entrance to a club. Everyone got in. The bouncer didn't allow me in. I had no drugs, nothing. This is denial of freedom of association. I told that to the police around there, but they just told me to go home. I felt bitter and humiliated. I've lived in many places. And I think these things are a question of conscience and good will.

Alba's testimony is that this is not a single case, but his continuous experience: 'It's so appalling—the situation here.' Nicole (36), another Berliner, described a somewhat different situation, also referring to social rather than formal exclusion. When asked about what aspects of inequality in the city come to mind, she argued that there were no particular groups who were excluded, but people who do not fit 'because they are not eccentric enough' feel awkward and not wanted in places. The quality of life in the city is when it is open, she argued. It means everyone can do what he or she wants

> even when you don't fit into the norm, or the cliché. (. . .) I think that's an important aspect of quality of life. (. . .) If you don't fit in, into the scene, e.g. fashion style-wise—when you're not hip enough or not sufficiently eccentric, for example, people won't consider you as equal.

This reminds us that even those who prize difference and eccentricity can have negative attitudes to those who are very conventional, and therefore challenges to relational equality can occur in unexpected places. Moreover, while in Alba's situation, the exclusion is based on social norms or prejudice, in Nicole's case, not only is the exclusion based on norms, but the excluding group is not even aware of the act of exclusion. Nicole does not feel welcome in certain neighbourhoods or shops even though those who cause this feeling are not aware of it and in fact, believe that they are open-minded and inclusive.

Such exclusion, often informal, is closely related to not feeling at ease in public or being unable to be oneself in public. It could be due to your accent which immediately reveals that 'you don't really belong', unless the city takes variety of accents for granted or even celebrates it. Ruben (26) from Amsterdam is originally from East Holland, which is often considered inferior in Amsterdam. He says:

> I think that the thing that most bothers me about the city is probably the way people judge me according to my eastern accent. It's impossible to hide my eastern accent when I speak Dutch. (. . .) [It is difficult because] language is the first hurdle you must jump over if you want to build trust.[5]

Sabine from Berlin tells us that the east-west divide is still very strong, and people from the west (*Wessies*) don't mix with people from the east (*Ossies*) and vice versa:

> People still have surprisingly little contact to each other. My colleague, who is from the East, tells me that she has never been to the Wannsee [a famous bathing resort and a beautiful lake within the city of Berlin]. I can't understand her . . . only because it's in the West??

Sabine points to another component of exclusion: for a city to be egalitarian people from different neighbourhoods should not think twice about whether to visit another neighbourhood and how they would be treated there.[6] R from London describes her own experience as an example of social exclusion:

> I work in Chelsea. People there talk down on you a little. Not everyone, but some. And this really depends on the area. (. . .) But, you know, it's just these things . . . like when you're walking into an area or a shop and don't feel welcome.

[5] At the same time, some Amsterdammers make it a principle to learn the language of the immigrants, because integration, they claim, is not a one-way process, whereby the newcomer is integrated in the dominant culture, but rather a process whereby each community, those originally from the city and the newcomers, learn each other's cultures and adopt parts of it. For this reason, some interviewees in Amsterdam called their city 'the milkshake city'. See de-Shalit (2018).

[6] While we endorse the idea that city dwellers should be able to go anywhere in the city without feeling they are unwelcome, we do acknowledge that occasionally cities can allocate spaces or buildings for the use of specific groups such as for women. In several cities, including Rio de Janeiro, Tokyo (on some suburban and train lines), Delhi, Mumbai, and elsewhere, there are special carriages on trains for women only. Such acts are done because of egalitarian reasons and are conducted with the blessing and agreement of the entire community, but at the same time are in a sense regrettable, as they are a response to the fact that many men behave in ways in which women find threatening, and therefore is itself a response to injustice.

Distance too can be a barrier. Sulaika (36) also from London describes a situation whereby

> London's community is quite subtle. (. . .) Poorer people are located away from a large part of things, from experiencing different things. So, they're at a major disadvantage.

Tid (30), a Londoner now, who we mentioned before in connection with 'working-class pride' in Liverpool where he grew up, describes how the disadvantaged were segregated spatially and neglected to the extent that

> the cliché vision is well either you join the army or you're on welfare from the government or you are in prison. The slightly ambitious outside option was to play football.

A city of equals should be inclusive in most if not all its neighbourhoods. Oded, 40, from Tel Aviv, works in the south of Tel Aviv where many asylum seekers and foreign workers reside. He loves it, especially the diversity and enjoys every minute of the meetings of cultures. However, when asked about equality in the city he is very clear:

> [it will happen only when] the diversity occurs across the whole city, and not just in the south. I want to see what happens in the north, where the houses are worth 5, maybe 10 million Shekels ($1.5–3m), when a group of Eritreans will stand in a doorway. It will take less than a minute for a city patrol to get rid of them.

He might be exaggerating, but nevertheless there is truth in his claim that people who live in north Tel Aviv can be very proud of how inclusive the city is but stop short of actually living with asylum seekers.

The extent to which asylum seekers and refugees are integrated is obviously a matter that affects inequality in the city. Claudia (58) from Berlin says:

> And what about refugees, what kind of inequality do you think of here? Well, that they cannot work here, although they are well educated. Germany laments a shortage of skilled labourers. And, nevertheless, people aren't allowed to work. Their degrees and professional qualifications are not being recognised here.

As we described in Chapter 1, with respect to immigrants, refugees, and asylum seekers, cities often promote more egalitarian policies than the states where they are located. For example, in October 2020, when the government of Italy announced that it would not allow a migrant-rescue ship approaching Italy to enter any Italian harbour, the mayor of Naples, Luigi de Magistris, challenged this decision, announcing his support of the organization ResQ Onlus in its plan to launch a migrant-rescue ship in the central Mediterranean. However, much remains to be done, and often refugees and asylum seekers who do enter the city are treated with disdain both by the authorities and other residents, and have access only to a limited range of the city's facilities.

4.2.3. Political Standing

Many of the elements of the city of equals we have discussed so far concern informal social structures—who feels welcome and who doesn't—or the general consequences of laws and policies as they affect different groups, such as the effects of encouraging cycling in the city. These policies can affect people in very individual ways that are not necessarily determined by their particular group membership. But there are also questions where very sharp decisions need to be made, as they need to be embodied in law, such as whether immigrants prior to naturalization have the right to vote in local elections or stand for office. Of course, there is also a less-formal aspect to these political questions, such as who will have the confidence to stand up and speak at, say, a planning meeting about urban renewal, and who will be taken seriously. In Chapter 2 we mentioned Cassiers and Kesteloot (2012) who argue that spatial segregation has three implications: diminished opportunities for income, for those who live far away from the economic centre of the city; lack of social network and social capital, which in turn make social mobility even harder; and stigmatization and lack of political representation, as those residing far from the centre and in more deprived neighbourhoods often receive less political attention. So it seems that the informal is linked to the formal, and results in lack of representation. As we also explained in Chapter 2, two theorists have shown this in detail. Loren King (2011) claims that a city's regulations and policies do not always take into account the values and norms of all groups in the city, or, as he defines it, whether it gives equal political standing to all who will be affected by the policy. And Patti Tamara Lenard (2013; 2015) describes representation in councils in cities where immigrants form a very large group, such as Toronto. Because immigrants tend to reside in the same neighbourhoods, and because many of them haven't yet naturalized, and because in Canada, those not yet naturalized lack voting rights in local elections, let alone national ones, the situation is quite bizarre. It is not only that the residents are not represented, but that the entire neighbourhood is not represented, especially when the elections to the councils are based on constituencies, which yields a situation whereby the neighbourhood, as a constituency, lacks representation.

Lack of representation matters in many ways. In terms of equality, one stark issue is renewal of infrastructure such as sewerage and drains, water supply, road and sidewalk surfacing, and tree-planting, where provision can be very uneven between neighbourhoods. The newer a neighbourhood is, the easier it is for the municipality to provide efficient infrastructure. As Omri (28) from Jerusalem says:

> It all comes down to the infrastructures. The good neighbourhoods are either the new ones or the renovated ones. Since the municipality doesn't really have the money to invest in renovation, the new neighbourhoods are always better than the older ones.

We would add that wealthier neighbourhoods, and those with better political representation, are also more likely to speak up loudly, and make it uncomfortable for city officials until their voices are heard and acted upon.

Hence although we agree, of course, with Patti Lenard that having a vote matters,[7] and is perhaps the most important factor, political participation includes other spheres of life, and having a vote is one parameter among many. Even representation itself is important not only in the municipality but also in the city's civil society, NGOs, and, of course, the media. In a city of equals a multiplicity of groups are represented in local media and radio, and the local TV broadcasting showcases the variety of local cultures and languages, thereby catering to the needs of all groups but also enabling people to enrich themselves by becoming familiar with the other in the city. Moreover, in a city of equals, people will not see such representation and participation as something special. They will, actually, be eager and open to listen to others. A good example for this was an interview with Yap (67) from Amsterdam. He says he does not have any problem with immigrants; actually, he adds, it's the other way around. Then he explains:

> Some people, mainly white males, think they are somehow entitled to be the guardians of the world's wealth. They are not. The fact that I was lucky to be born here [as a white Amsterdamer] does not mean that the ones who are less fortunate

[7] Lenard's thesis is supported by Frug (2011) who challenges US voting laws because they enable some city-zens, for example, long-standing residents, to vote and decide for others who lack political power and voice.

> are not entitled to the same opportunities. I would hate it if the city would not accept immigrants.

He adds, 'What's mine is theirs.'

For this reason, many interviewees mentioned that representation on all levels and in many spheres should be open to everybody. While it is true that in many cities there are discernible xenophobic sentiments, it is also true that in many cases a large proportion of residents believe that a healthy and egalitarian city includes representation and participation in as many spheres as possible. As Sherry Arnstein argued, there are different levels of participation, some of which are no more than lip service, or, as she puts it, a token. These are informing, consultation, and placation. The municipality sends warm messages of inclusion which in practice amount to little and act as a cover for inaction. But the main point for us, here, is that participation not only comes in different levels, but also in different forms (Arnstein 1969).

In the interviews people talked less about their voting and more about accessibility to decision-makers and bureaucrats and about being listened to. Moti (61) from Tel Aviv complained that not all residents of the city have equal access to the decision-makers at the municipality, and so did Mano (70) in Jerusalem, who thought that this is not necessarily a matter of inequality, but rather something that affects everyone, whatever their situation—people are not being listened to, so it creates a different kind of political inequality, between those in power and all the rest. On the other hand, Tina (49) from Hamburg suggests that the problem is especially acute for homeless people who really lack access to the bureaucrats. It seems clear that there are various levels of not being listened to and being listened to. A person might lack the appropriate skills and knowledge and simply have no idea how to approach the municipality. Or they might have the skills, but despair about the prospect of ever being listened to, so refrain from even trying. Or they might try but fail to reach the right audience. Or they might be listened to but ignored, and the authorities still do whatever they want, as their listening is not more than a token. Or they might be listened to, and taken seriously, but their views ultimately have no influence as, for example, local politics is all about the old boys' network, and only some voices have impact. These are all issues that affect whether one's city is a city of equals.

Political standing is not only about having your concern being assessed by a legitimate, fair, and open (transparent) process. Rather city-zens also want to win from time to time. A city of equals pays attention to the interests of a variety of groups according to whatever criteria the group is formed, and does what it can to ensure that those interests are met; not that they are always met, and not that they are never met. Even the language used to refer to people can make a huge difference. In Canada it is common to avoid the term 'refugees' and use 'newcomers' instead, sending a powerful message of inclusion.

The general idea of political standing is worth further reflection. It relates first and foremost to how decisions are taken by the city's authorities. If those in power wish to be egalitarian in their attitude to the city's population, they should see that no particular group regularly finds itself constantly on the losing side unless their demands are unacceptable because, for example, they are utterly fanciful or harmful to others. The principle should also apply to the city's civil-society institutions: school boards, local NGOs, local boards that run citizens' initiatives such as communal gardens, and similar. Nobody should feel alienated because they repeatedly find themselves in the minority. We acknowledge that there is a price to pay here because, to achieve this, sometimes the majority will have to sacrifice its own idea of the good or decide against its interest. But the benefit of what might be termed political inclusion and refraining from alienating city members is of great value.

When things go well in a city, people express great civil pride. Heidi (Hamburg) says: 'There is little that I dislike or could complain about. I am a Hamburger in heart and soul.' But there can also be estrangement. Christian (40) is in Berlin but used to live in Hamburg. He says: 'Making contact in Berlin is very difficult. (. . .) It would be nice to have a social climate (. . .) In Hamburg, for example, people are much more approachable then here, I feel.' Claudia (57) regrets that Berlin's homeless people do not feel at home enough to participate politically, and as a result they, so to speak, do not co-own the city.

Renata, a psychologist in Rio de Janeiro who lives in one of the affluent neighbourhoods, frankly says: 'I think I got used to seeing this inequality, and I try to protect myself.' So yes, many city dwellers can live side by side with inequality as long as they are not harmed. If due to inequality there is a high crime rate, as in Rio, and many other cities, they try to 'protect themselves.' If they succeed, they have little direct motivation to contribute towards reducing the inequality they see around themselves. Therefore, an egalitarian city should see that people do not retreat entirely to their comfort zone, by which we do not mean that people should voluntarily expose themselves to crime, but should be aware of their privilege and consider how the lives of others can be brought to the same level. Lucia (42, New York) mentions this when she says that people believe that state and city services for poorer people are 'favours rather than rights', which is a very important observation, as, we will show in Chapter 5.

4.3. Conclusion

In this, and the previous, chapter we have drawn out a number of themes from the interviews, and although we have organized them under our key values, there is more work to do to put them into a firm theoretical framework. Drawing on the interviews, the literature review, and our own reflections, articulating such a framework is the task of Chapter 5. At this stage, our preliminary conclusion is that the interviewees have shown that they are greatly concerned not only about their own fate in the city, but its overall character. Few gave any indication that they thought their cities had done too much for other groups, except for the rich and privileged. There was a concern that the people who are the lifeblood of the city—service workers both in the public and private sector—as well as members of minoritized groups and newcomers, were not always treated fairly in terms of access to facilities and resources. Areas of particular concern were, naturally, housing, transport, and access to leisure facilities. But equally there were enormous concerns expressed about how people treat and relate to each other, especially in street-level interactions. Constructing a city of equals is a highly complex task, which might not be reducible to an easily applied formula. We will give our account of what it means more specifically in Chapter 5.

Before we continue, we should address a possible challenge. We have described many themes; but when cities design their policies, they might face the problem that catering for one of them could make another less available. How do we recommend such tensions should be handled? We ask the reader to bear with us. In Chapter 5 we will first give a fuller picture of how these themes can be grouped into four core values, and then discuss how to approach potential tensions.

5
A Secure Sense of Place

5.1. Introduction

In the broadest sense, two types of concerns recurred in the interviews: the type of facilities, services, and resources people had access to, and how they felt treated, whether by other city-zens or by the civic authorities. Putting these two together leads us to our key claim. When city-zens reflect about what a city of equal means to them they think about feeling welcome or at home in their city, and in particular having a sense of belonging, which can be broken down into several general core values which we have mentioned several times before but will now develop in detail. Accordingly, we argue that in a city of equals everybody has a secure sense of place. But what is interesting, and this became clear in the interviews, is that when city-zens think about a city of equals, they don't think only about themselves. People want to be accepted for themselves and in their own right, *although among others with similar entitlements*. That is the egalitarian sentiment of the city. No one settled in the city wants to be regarded as a guest or a servant or an outsider or as someone visiting or passing through, or merely tolerated but not welcomed. Rather they wish to be considered to be someone who is recognized as having as much right to live within the city as anyone else, and who makes a contribution to its richness and vitality. At the same time, they understand that a city of equals implies that this experience of recognition should extend to all, including, but not limited to, newcomers, immigrants, the young, the elderly, people with disabilities, those who identify as LGTBQ+, all ethnic and racial groups, and people from all social classes.

As indicated, drilling down further, and based on the material we explored in Chapters 3 and 4, we find that the idea of a secure sense of place can be understood in terms of four core values. These overlap to some degree, but are worth emphasizing independently:

1. Accessibility to the city's services is not constituted by the market.
2. A sense of a meaningful life.
3. Diversity and social mixing.

City of Equals. Jonathan Wolff and Avner de-Shalit, Oxford University Press. © Jonathan Wolff and Avner de-Shalit (2023).
DOI: 10.1093/oso/9780198894735.003.0005

4. Non-deferential inclusion (that is, being included without having to defer).

We arrived at this conception by developing and refining our initial understanding of a city of equals in the light of our interview results, as we explained in Chapter 1 and elaborated in Chapters 3 and 4. Indeed, we will continue to draw on the interviews as we illustrate the themes, as well as the literature review from Chapter 2 and some further sources. We will begin by discussing the sense of place and belonging, and then will consider each core value in turn.

5.1.1. The Sense of Place

Urban life and activities are often related to space and a sense of space. As sociologist Jennifer Cross beautifully explains (Cross 2021), a sense of place is an inter-disciplinary concept, and therefore, to grasp its full meaning and complexity we need to understand how anthropologists, psychologists, geographers, landscape architects, sociologists, and—we would add—philosophers, perceive a sense of place. This is consistent with what we saw in Chapter 2, where we discussed theorists from many disciplines who wish to understand inequality in the city largely in spatial terms. What seems to be common to all accounts is that a person's sense of place involves particular experiences in a particular setting or space, which are interpreted via culturally shared values or beliefs and sometimes practices, and which refer to both the physical and social environments of the person. All these result in attributing some culturally shared *meanings* to a particular space and in some bonding between the person and the place. Sometimes—when the sense of place is very strong—these also result in the person asserting that the place constitutes part of their identity—who they are—thereby defining a kind of attachment to the place. This attachment can involve biographical dimensions, most commonly a place where one was born and grew up, but also where one fell in love for the first time, or where one went to university and stayed on, or found work after a long job search, and so on. Equally, it can involve community or heritage dimensions, such as the place where our first national institutions were established, or where we stopped the enemy's army from advancing, or where our ancestors had the first open-air theatre, among many other possibilities. More controversially some may think that a sense of place could also be associated with negative memories; a battlefield or the site of an atrocity, which could be even more difficult to reconcile with a positive sense of identity if one's forbears were among the perpetrators.

Whatever we think of the negative aspects of place Cannavò (2007, 20–1) captures its positive form well: Place is not just an object. The creation and identification of something as a defined place, he argues, is a process of social construction. Thus, place is an essential human practice. It involves the physical and conceptual organization of our surroundings into a coherent, enduring landscape. Indeed, we can imagine a person, walking and wandering through the cityscape, thereby constructing

> a very concrete and vivid image loaded with meanings and emotional attachments to the street life, local cafés, grocery shops, schools and all the fundamental components of quotidian urban life.
>
> **(Barak and de-Shalit 2021; see also Löw 2013)**

It is interesting to note that when people interpret the city in which they live as, for example, 'beautiful', 'vibrant', 'exciting', 'relaxed', 'cosy', 'friendly', 'old fashioned', and so on, often they are ascribing a meaning to the place. Of course, they do not do this in a vacuum. They reach a sense of place through learning, incorporating, and understanding the way their neighbours, community, even the city itself understand the place. And when people reach this sense of place, they feel that this is *their* place, that they feel at home, they belong, and hence can flourish and develop in this place. The point is that this is very much a contextual, iterative, process. As Bart van Leeuwen writes (2010, 2018), the patterns of recognition that people live with define a kind of space in which one is able (or not) to develop a way of life that is truly human. Van Leeuwen here captures the essence of our idea of a sense of place.

The concept of a sense of place can helpfully be illuminated further within the context of the Capability Approach (Sen 1992; 2009; Nussbaum 2000). Arguably, having a sense of place is closely related to what Martha Nussbaum defines in her list of capabilities as 'affiliation', but in our case here applied to a particular place. A capability is a person's opportunity to achieve a functioning, and a functioning is something which people have a good reason to do or be. For a number of reasons, unlike Nussbaum, we often prefer to focus on functionings rather than capabilities, though we can leave that debate aside for present purposes (Nussbaum 2011; Wolff and de-Shalit 2007; Wolff and de-Shalit 2013). Nussbaum describes affiliation as:

> Being able to live with and towards others, to recognize and show concern for other human beings, to engage in various forms of social interaction, having the social bases of self-respect and non-humiliation. Not being discriminated against on the basis of gender, religion, race, ethnicity, and the like.
>
> **(Nussbaum 2000, 79)**

There is, however, an important difference between 'affiliation' and 'having a sense of place', which is in one way narrower and one way broader than affiliation as Nussbaum understands it. The narrowing of our conception, or perhaps better to say further specification, is that whereas affiliation is about belonging in the context of relations to other people, but without geographical connection or restriction, a sense of place focuses on belonging in a particular space, describing the relationship with artefacts located within it (buildings, streets, pavements, parks, etc.) as well as the people who use them. In this way, then, the set of relations with others that partly constitute a sense of place is narrower than those that generate affiliation in Nussbaum's sense, which will include connection to friends, family members, and others one has a connection with, living a distance away.

This restriction brings with it a broadening, which may be implicit in Nussbaum's account of affiliation, but which we need to bring out explicitly: inevitably this spatial feeling is also social. For example, one's relationship with the street on which one lives includes how secure one feels in this street, and this, in turn, and as Jacobs taught us, derives from many factors, including: the architecture of the street's buildings (for example, are the buildings huge and apparently empty or built on a human scale and lively, are there cafes and shops that are open in the evenings?); the way people use the buildings and the streets and pavements (for example, do people tend to hang around, chat outside for long periods, do children play on the street?); how they behave (for example, do people shout and quarrel in the streets or do they respect each other?); their norms (for example, do people ignore each other or do they tend to assist you if you fall and twist your ankle, or if they know you are ill at home?), and also the way people think about how safe the street is.

Now, while in much social psychology having a sense of place is conceived as part of one's well-being (e.g. Hernandez et al. 2007; Cresswell 2009), we argue that it is also a functioning. This is because this sense of place, this understanding of one's environment, is a key component of one's identity. It attaches the person to something which is greater than just his or her individual self. One is not just a person, one is a New Yorker, or Shanghainese, or Istanbulie, or a Porteño (the name for residents of Buenos Aires) or a Carioca (the name for residents of Rio de Janeiro), or a Parisian.

Not only is sense of place a functioning, it is also important in that it is a 'fertile functioning', by which we mean that it strengthens other functionings or makes them less insecure (Wolff and de-Shalit 2007). As many scholars have found when researching immigration and integration (van Bochove 2012; Okamoto and Ebert 2016), feeling at home and having a sense of place affects behaviour in a positive way. It makes people feel more self-assured, more positive, and capable of doing more things. To see this, think of yourself in different neighbourhoods or streets—some in which you feel a stranger, and some in which you feel at home—and imagine how you behave in each one of them. These feelings affect the entire range of our everyday activities and they have an impact on many of our cognitive and emotional capabilities.

Moreover, when people have a stable and strong sense of place they often become more tolerant of and open to others, including immigrants. Similarly, people who lack a sense of place often end up not tolerating others, and in particular not tolerating immigrants, whom they perceive as a threat to their already unstable and vulnerable sense of place. Given that in cities people meet others, including immigrants, on a regular basis, this fertile functioning becomes fertile for *everybody's* sense of place.[1]

Therefore, no wonder that having a sense of place is sometimes seen as a strong interest. The philosopher Margaret Moore (2019) argues that people have place-related interests which derive from the role of place in their life plans and their life projects. In fact, she argues, a connection to a place and a sense of place is an important precondition for people's projects, plans, and way of life. She claims that people live their lives and make choices and decisions against a background context; and that land or place is such a context, which people assume as part of the fabric of their lives, from which they make choices and pursue their way of life. As sociologist Göran Therborn (2009) argues, places 'mold actors, structure their life chances, and provide them with identities and traditions of social and political action' (Therborn 2009, 529).

However, a sense of place can become insecure. During wars, or violent clashes, city-zens' attachment to a place could start to become under threat or more questionable, especially if it is destroyed, as we have recently seen in Ukraine. The point is not that people's psychological attachment falters (although it could do), but that they become anxious about their prospects of flourishing, or even surviving, there. Similarly, victims of environmental disasters, floods, or droughts report that they fear that they might not be able to remain in their homes, or to cultivate their land, which we interpret as a way of expressing that their sense of place has become insecure. They may well still have a sense of place, but it is becoming less secure, compared to somebody who lives in a place which is not in a war or is not vulnerable to repeated flooding.

A person's sense of place can also become insecure if others around them send them a message that they are no longer welcome there. Jews in Berlin

[1] We thank Bart van Leeuwen for this point, as well as for other insightful suggestions.

at the beginning of the twentieth century had a very strong sense of place; they enjoyed more political and economic rights than probably most of the Jews in Europe had ever done before, and felt not only that Berlin constituted their identity but also that they were an important part of the body of city-zens who constituted Berlin's unique character, as they were very much involved in Berlin's spheres of art, literature, philosophy, and science. However, in the early 1930s when the Nazis rose to power, this sense of place became insecure, as was described by the novelist Stefan Zweig in his memoir *The World of Yesterday*. Zweig (1943), albeit in relation to Vienna rather than Berlin, describes how in no time the Jewish intellectual elite of Vienna, which had a very strong sense of place in the city, found themselves utterly bewildered, feeling that they could not recognize the very Vienna which they had loved so much. Zweig himself was one of hundreds of thousands of Jews who left Germany and Austria, in Zweig's case travelling ultimately to Brazil, where soon after arrival he and his wife tragically took their own lives. Other German, Austrian, and Czech Jews who, for whatever reason, decided at first not to leave, later found it too late to arrange visas, but in some cases were able to send their children on the Kindertransport trains as refugees to the United Kingdom, hoping to join them later, which some, but far from all, managed to do. The place which was home became a threat to life.

Of course, the examples of Berlin's or Vienna's Jews in the 1930 are very extreme, and no hostility or exclusion in today's world is on the same scale and level as the Holocaust, although many minoritized groups, families, and individuals continue to face similarly tragic fates all over the world. Some of our interviewees had been under such threats themselves and provided stark illustration of the idea that without a *secure* sense of place it can feel that one simply doesn't belong. Ibrahim, 36, is originally from Syria and had lived in Turkey, Sudan, Libya, and Italy before moving to Hamburg where we interviewed him. He says that his hope is that one day it will be safe for him to return to Syria, but he quickly adds that Hamburg is like a second *Heimat*, a word meaning home in the spatial sense, in a nostalgic, and both biographical and national sense of the word. For him, he says, this means that even though he is alone in Hamburg, having arrived only recently, everything seems fine. But at the same time others in Hamburg have been made to seem something of an outsider. Jason, the young Black German construction worker in Hamburg who we quoted before says:

> You know . . . I have been to Stuttgart, to Munich . . . and I was being looked at strangely, because I'm Black. In Hamburg, and I think also in Berlin, we accept everyone, and it's truly multi-cultural.

But even so, as we discussed in Chapter 4, Jason adds how at work some people looked at him suspiciously because they realized he was a Muslim, and this made him realize he was not part of the place in the same way as everybody else. Marzieh, 25 years old, originally from Tehran, now in Hamburg complains:

> There are always weird or crazy people, everywhere . . . For instance, people talking about my headscarf. They come straight to me and ask me why I wear it. I grew up like this, and perhaps I don't have sufficient information and opinion about this but I think it's a question only for those who wear it.

But she reassures us as well that she has never experienced inequality in treatment in official matters.

Some non-Muslims in Hamburg may worry that if Ibrahim, Jason, and Marzieh feel at home there, with a good sense of place, it will put their own sense of place at risk. And indeed, it may be that it is common to believe that enabling some people to find a sense of place implies that others will have less of it. In academic terms, such people believe that a sense of place is a rivalrous good, a zero-sum game. And yet, we want to argue, having a sense of place can be seen as a non-rivalrous good too. Enabling a sense of place to immigrants can be done without compromising the sense of place that exists among the city's more long-standing inhabitants; similarly, a city can enable a sense of place for elderly people without compromising the sense of place that exists among its younger city-zens, and so on, even though there are also ways in which conflicts can emerge and compromise is necessary, such as over issues such as noise in the streets late at night.

At this point, we should concede that seeing a sense of place as a non-rivalrous good might appear to some readers as too naive and optimistic, as 'easier said than done' because empirically, in some neighbourhoods locals see immigrants, or for that matter gentrifiers too, coming and changing the ethnic or class composition of the neighbourhood, and this weakens their sense of place. They claim that their new neighbours do not speak their language, or share their styles of dress, food, or celebration, or that they seek very different communal ties from the ones that are already there. While, as we have noted, multiculturalism appealed to many of the interviewees, we do not claim to put forward any accurate empirical generalization of how all, or even a majority, of city dwellers think about newcomers to their neighbourhoods. Indeed, as we explained in Chapter 1, we terminated some interviews early when racist opinions were expressed, for our task is to construct a theory of a city of equals, not to conduct a survey. So let us clarify: we are not treating every voice in the city as equally valid for our project, even though we accept that there may be a superficial tension with our argument for equal political standing for all. But the tension disappears when it is realized that our project here—as distinct from our views about political representation in the city—is only to give voice to those that will help create and deepen a normative theory of a city of equals. So we remind the reader that by claiming that a sense of place can be non-rivalrous, we do not suggest that this is so with every city dweller worldwide.

Cities can potentially enable us to have this sense of place due to their very nature: they are systems which work to connect, create bonds, and form attachments. This happens at the workplace, in sports events, at rock concerts, on the street, and at the small park around the corner when one meets neighbours, when city dwellers walk their children to school, or walk their dogs, or sit with friends for a drink, shop, or just stroll in the street. Cities and the communities which live in them bind city dwellers to their history, personal and collective memory, language, natural surroundings; to things with which they are familiar and at ease.

Yet at the same time inequality can enter the picture. Cities might intentionally or unintentionally harm or risk the sense of place of some of its inhabitants. Which policies sustain this sense of place and which weaken it among some groups in the city needs much more empirical research, as to the best of our knowledge only a small literature exists on this question (Feitelson 1991; Devine-Wright 2013; Žlender and Gemin 2020). Nevertheless ideas which we frame as working through the questions of who does, and who does not, have a secure sense of place were at the centre of many of the interviews we discussed in Chapters 3 and 4. One poignant example from the interviews of how and when the sense of place can unintentionally, but thoughtlessly be disturbed among some of the city-zens, concerns policies of street naming. Cities often name streets to consolidate the city's collective memory by linking to notable people or events evocative of local or national symbolism. This is part of what offers sense of place to the city dwellers. However, in Jerusalem, for example, there are Arabs who live in the western (predominantly Jewish) part of the city, in streets named after the pre-independence Jewish military organizations (Etzel and Haganah). Nona, a Jewish resident sees this as highly problematic, saying:

> The fact that they [Jerusalem's Arabs] have to live in a street that is called 'Etzel' is a major bad. Moreover, there is nothing they can do about it. They can't say 'please change the name of the streets'.

A sense of place is also related to trust. People consider a sense of trust, if it exists in their city, as precious. Mano, 70 years old, lives and work in central Jerusalem and recalls from his childhood:

> When my mother used to go to the market, say to buy tomatoes, she wouldn't ask 'how much does it cost?' She would just take a note, put it in the hands of the merchant, and trust he would give her back the right change.

Hany from Tel Aviv claims that familiarity and trust have changed over the decades:

> When I was a child, you could walk four or five streets away from your home, walk into any apartment, and feel like it is your own family . . . I have been living in the same building for thirty-five years, and now I don't even know the names of all the residents in the building. They won't talk to me. This is not privacy; this is snobbishness. The municipality has to be blamed; they have been developing the city into something snobbish.

Resonating with the concerns of other interviewees we discussed in Chapter 3, Hany blames rising rent prices for a process of gentrification during which long-standing residents lost their trust in their neighbours, hence their sense of place.

In the city the sense of place includes the opportunity to be amused, entertained, and even astounded or amazed by the things around us. We go to parks and we see the children running, and people trying out their football skills and we are impressed with their vitality and talent; we listen to a woman singing, accompanying herself with her guitar, when we take the tube. Even window shopping for goods we could not dream of owning contributes to our experience of the city. The richness of life is connected to seeing what other human beings are capable of doing: building a beautiful cathedral with what must have been limited tools, crafting the treasures in a museum, and designing and building breathtakingly imaginative skyscrapers. We ask ourselves how could humans make and build such things? Who thought of putting those ingredients together to create such a great meal at a restaurant? We go to a cafe where the barista is professional and we treat ourselves to a really good espresso, especially on a cold rainy day. We enter a good book shop or library and think—how wonderful it would be to spend the next month just sitting here reading. In the city we have so many opportunities to be astounded with what humans can do, marvelling at human achievements and excellence. Many theories of the human good rightly emphasize the importance of being creative. In doing so they sometimes neglect the other side of creativity; how life-fulfilling it is (for both sides of the relationship) for there to be an audience to enjoy the creations of others.

But not everybody has this opportunity to be amazed and astounded. Nathan is 18 years old, a young man, born and raised in Rio de Janeiro. He has black roots and is the father of a two-year-old child. He lives in one of the favelas with his mother together with two brothers and three sisters. His son does not live with him, but with his ex-girlfriend. He stopped schooling early. He complains about 'lack of opportunities to study, teachers who are authoritarian at school . . . It is a lot of shit.' He describes how because he had to help at home he could not study. But what made a difficult situation impossible was the city closing down the CIEPS, the integrated centres of public education, where he had studied. These were established in Rio de Janeiro's poor neighbourhoods, where high-quality public education was offered to everyone. But, as Nathan complains, 'Bolsonaro and Witzel[2] are both destroying us here'. He adds that if he had free public transportation he would be able to see the city. Both Nathan and Valentina (37), also from Rio, complained that because the police are violent towards the poor, they are afraid of even trying to reach places in the city, and so cannot enjoy what Valentina describes as 'the outdoor life of the city, where we get to experience the Brazilian culture, including bars, music (especially samba and funk) and dance performances'. Arthur (35) from Rio says that 'the poor are being wiped out here'. He adds:

> I'm from a privileged group. I'm white and well-educated. A black person, on the other hand, is someone who will be less regarded in Rio by other citizens. Think about how people choose to sit inside the bus? It is very often far away from a black person. I also notice that policemen treat a black person very violently. This is taken for granted by people. The blacks here are used to being segregated and discriminated against, and the whites are used to treating them badly.

Who has the opportunities to be amazed and astounded will depend, in part at least, on where and how you live. As discussed in Chapter 4, Jane Jacobs advocated local small parks, and it's easy to understand why, but if that's all we have we miss the big astonishing park, with landscaping, trees, and large lawns acting as impromptu playing fields, often for improvised unusual team sports. New York without Central Park is unthinkable, even if the number of small local parks increased. Yet we should also bear in mind Margaret Kohn's

[2] Wilson Witzel was Rio de Janeiro's governor at the time of the interview. He is from the same party as Jair Bolsonaro, Brazil's populist president at that time; his policy emphasized security more than anything else.

observation mentioned in Chapter 3 that a highly regulated park will be less attractive to those who want to engage in boisterous, loud, and exuberant activities; indeed, to those who like to express their enjoyment. But still, if in a city there is a culture whereby it is common for people from various neighbourhoods to move around, to see other neighbourhoods and therefore to have more chances to be astonished, and in different ways, then it is more egalitarian than a city where there is no such culture because of, for example, racism or classism, because of narrow horizons or just lack of convenient transport.

As noted above, we have found it helpful to break down a healthy sense of place into four themes that slightly overlap but can be analytically distinguished, and which we will call 'core values'. We approached the interviews with a general conception in mind, which we greatly enriched through the interview methodology. But the interviews don't speak for themselves. We need to select and collate the materials and put them into a compelling order, not unlike the work of a good film editor, taking scenes and drawing out a narrative, which we attempt in the following, while accepting that it is possible that there are other ways of presenting our results.

5.2. The Four Core Values

5.2.1. Core Value 1: Access to the City's Services Is Not Constituted by the Market

One form of significant inequality in some cities is unequal access to various services and amenities of the city. This includes what we termed in Chapters 3 and 4 basic services such as food shops, markets, pharmacies and health clinics, or schools for children with special needs; leisure services such as museums, cinemas, pubs, cafes, parks, playgrounds, tennis courts and football grounds; and public amenities, from public toilets to libraries. If only the wealthy or the ethnic majority group can live in convenient locations, easily reach the beach, listen to live music, or find somewhere to play tennis, or even meet with their friends, if they enjoy more cafes and pubs per person than the poor or the groups that are minoritized, if they find food markets and farmers' markets, health food shops, clinics, psychological services, and the like within easy reach and walkable distance, while the poor or the minoritized do not, then there is a very visible form of exclusion in the city, which, perhaps even just as importantly, gives those excluded diminished opportunities for enjoyment. It also, centrally for us, links to a sense of belonging and sense of place through the idea of exclusion: a city dweller might feel that: 'The city is not for the benefit of the likes of me, but only the wealthy and privileged. If I cannot make use of the things the city is famous for, I am not part of its story.'

If, on the other hand, the city provides everyone with access to valued facilities, it no longer suffers from this defect. Of course, in liberal countries with a market economy wealthier people are inevitably more likely to be able to find more luxurious places to live, and in some societies can also purchase preferential medical care and education for their children. They will also be able to find exclusive forms of access to beach or tennis clubs, or reserve choice seats at the concert hall. But if there's subsidized public housing or rent control, universal health care and education, an even spread of clinics, pharmacies, and medical centres so that nobody has to travel far to reach them, then there are forms of inclusion to basic and other services for all. Also, it is important that these services are of good quality, especially concerning education as we discussed earlier. In some megacities, residents of poor neighbourhoods do not need to travel far to reach a school or a college, so prima facie, it appears as if the accessibility of education services is independent of the market. However, to reach a *good* school, which are usually located in the city centre or the more affluent neighbourhoods, they do have a long and exhausting commute. When the quality of the school or college is a factor, then the picture changes (Giannotti and Logiodice 2023).

Similarly, if there is a public park or beach open to all, public tennis courts rentable by the hour at reasonable rates, ways of lining up for a cheap theatre ticket on the day, or free concerts in the park, free entry to museums and libraries, and so on, everyone can feel that the city's facilities are their own too. As Renata from Rio de Janeiro says when she reflects about what gives her a good feeling in the city:

> I really like to go to the beaches here . . . enjoy the urban nature areas during my weekends—such as walking around the park 'Lage', Flamengo Park, and Morro da Urca. Such beautiful natural spaces are open to all. You don't need to pay to enjoy them.

This is, then, essentially a matter of non-market access to goods: public, concessionary, or collective provision of goods and services, so that one's general success in urban life does not track one's economic success, or class or other circumstances. In an unpublished paper Thad Williamson presents, as a definition of 'mobility' as 'the capacity to access and participate in the economic, civic, and social life of one's community, without respect to social class or personal economic circumstances' (Williamson 2013, 1). Whether or not it is an appropriate definition of mobility, it provides one way of understanding what we mean by this present theme, although our emphasis here is less on economic life (important though it is) and more on civic and, especially, social life. The city has much more to offer than economic and political participation, for example participation in projects of preservation and conservation (Light 2003), and we are especially interested in those activities that make up the rich fabric of people's informal existence and experience.

Public services and facilities provide material benefits, but also strengthen the other categories we will come to below, such as social meaning, and non-deferential inclusion. In some cities there is very little public or collective provision of goods and services, and hence one's general success in life closely tracks one's economic success. But this is unfortunate in placing so much emphasis on income and wealth, especially when, very often, access to financial resources is only partially under individual control. With the right public support, virtually everyone has the chance of a life that is successful in some way, especially in the sense that they can make use of the city's potential, even if with some limits to options, whether or not they are economically successful. So alongside a strong economy and fair equality of opportunity a city of equals needs to have such things as high-quality subsidized housing, whether publicly or privately owned and run; an education system from kindergarten to high schools and colleges; transport, which is frequent, varied, inexpensive, and accessible; health care services either in small clinics or in hospitals; and parks, museums, and libraries which are either free or cheap; as well as a good physical environment.

In our interviews ideas related to non-market access to services came out in various, imaginative, ways, including creating the conditions that will allow individuals access to the labour market. In Chapter 3 we saw that many interviewees were concerned about play facilities for their children. Children also figured in another way. Andrea from Berlin (aged 47) and Erika, aged 75 from Hamburg, both claim that if they were mayor they would first and foremost provide free day care for children, so that their parents can both work, should they wish to. However, as we touched on before, in one memorable contribution, Erika adds also night care—a topic that seems rarely discussed in social policy—for parents who work night shifts, such as hospital nurses, people who clean up the city during nights, workers in bakeries, truck drivers, and so on.

We should comment, though, that market-pricing is not the only barrier. Access to goods can be blocked for non-financial reasons, such as travel time or limited opening hours, which can be particularly difficult for those hard pressed through work. For example, public parks, which are generally free to all visitors and publicly funded, are not necessarily accessible to all city dwellers, as their accessibility depends on how far away one lives (this relates to the third core value below, social mixing). Similarly, housing, and many other services, does not always have to be non-market to be affordable, and, vice versa, for even goods and services that are not supplied on the free market, most notably social housing, can be unaffordable for many. For example, social housing in Greece, or Germany, or Denmark can be expensive and sometimes comparable to market-rate housing.[3] This can happen also in cities where the rent-setting mechanism for social and affordable housing is neither income-based (rents are set relative to the assessed means to pay of an household) nor cost-based (rent are set at a level which allows the social provider to meet the costs of provision, including maintenance, insurance, etc.), but rather market-based: rents are set relative to 'market' rents, for example 80 per cent of the private market price. In such cases decent accommodation is not accessible for the significant percentage for whom it is not economically affordable. And even if affordable, long waiting lists, as is the case in Amsterdam, Barcelona, and other cities put public housing out of reach for many.[4] In this case social housing is non-market, but it is not really accessible.

Therefore, we wish to make two points. First, to clarify that by non-market accessibility we mean affordable and adequate provision of (high quality) goods and services, instead of allowing the rationale of the profit-seeking market to exert its effects. Second, that it is not the case that services *must* be non-market for the city to be a city of equals; it is just that we see that *in general* if the services and good are provided by the market then they are likely to be inaccessible and unaffordable to many people, with, for example, powerful groups lobbying for the end of rent control. Of course, this is an empirical observation and in cities where rents are historically lower, relative to income, financial barriers to access may present much less of a problem.

[3] In the EU in 2020, 12.3 per cent of the population in cities lived in a household where total housing costs represent more than 40 per cent of the household's disposable income (what is called the housing cost overburden rate). The corresponding rate for rural areas was 7 per cent. The highest housing cost overburden rates in cities were observed in Greece (36.9 per cent), Germany (22.2 per cent), and Denmark (20.3 per cent) (Eurostat n.d.; European Commission n.d.).

[4] According to a joint statement to the United Nations issued in 2018 by the cities of Amsterdam, Barcelona, London, Montreal, Montevideo, New York, and Paris, citizens' rights to affordable housing was at risk due to the influence of investors and mass tourism (among other things, Airbnb) on urban property markets (Cities for Adequate Housing 2018).

Although we are interested in access to leisure and cultural facilities too, as we set out in detail in Chapter 3, many of our interviewees, rightly, saw housing as the central issue. In many cities we heard complaints about the great expense of housing in more appealing neighbourhoods and districts. People would like to live in this or that neighbourhood but they cannot afford to. This complaint was bitterly expressed time and again in Berlin and Tel Aviv for example, which in recent decades have seen a major rise in housing costs which has made some previously accessible neighbourhoods out of reach for many residents, especially young couples or young families.[5]

One of the challenges any attractive city needs to face is how to balance the interests of residents against those of visitors to the city, who not only would enjoy the experience and bring welcome commercial activity, but also feel they have their own rights to freedom of movement. Some cities have taken a pragmatic approach of at least trying to regulate the flow of visitors. Venice, which particularly suffers from this dilemma, has introduced a new policy commencing in January 2023 whereby day trippers to the city will need to reserve their visit and pay a fee for their visit (NPR 2022). Many of Amsterdam's city-zens, supported by the municipality and the mayor, Femke Halsema, have been campaigning to reduce the number of tourists every year, and to ban tourists from buying marijuana in the city's coffee shops (Froyd 2021). Prima facie a city should be happy with a large number of tourists, as it brings income and is good for business. Business owners pay higher city taxes than residents, so once again it seems reasonable for the city not to try and reduce the number of tourists. But when Julia (29), a young Amsterdammer explains the rationale, she talks about the city's obligations to its residents, which override market considerations:

> I love the fact that Amsterdam is a global city, I'm proud of it. That is why most of the time I don't really notice that people talk to me almost exclusively in English. (…) [But] the city owes me more. If for example something happens now and I need to call the police or to call an ambulance, it would take much longer than it used to because we are now sharing these sorts of facilities with millions a year. The city owes me more than it owes you [i.e. the interviewer who was only spending a short time in Amsterdam], and yet it will take the police much more time to get to me.[6]

[5] From 2012 to 2022 the price per square meter in Berlin has increased by around 10 per cent per year, and therefore has much more than doubled when price rises are compounded, for both older and new homes (Yujelevski 2023).

[6] The question of a city's obligations to non-residents versus its obligations to its residents is not exhausted by this short discussion here; but, as we wrote in the Introduction, we cannot cover in a single book our question together with the question of justice between cities, or between a city and its environment.

5.2.2. Core Value 2: A Sense of Meaning and Meaningful Urban Life

As we noted in Chapter 3, from the outset one of our theoretical points was confirmed indirectly by the interviewees. When asked about what a city of equals would mean, only a few people mentioned income gaps in the city as a matter that concerned them in particular. On the other hand, many mentioned issues which we frame as 'having a sense of meaningful urban life', which, as we have said is consistent with Martina Löw's argument that city-zens regard their cities as 'entities of meaning' (Löw 2012; 2013). The type of attachment people have to their cities is based on their rich tapestry of experience, rather than reducible to a financial calculation.

In *The Adventure of the Copper Beeches*, one of Arthur Conan Doyle's Sherlock Holmes stories, Holmes turns to his friend Watson and says:

> It is my belief, Watson, founded upon my experience, that the lowest and vilest alleys in London do not present a more dreadful record of sin than does the smiling and beautiful countryside. (Doyle 1892, 277)

And yet, most city dwellers who read this will probably smile, thinking that Holmes is really exaggerating. Most city dwellers associate country life with 'nothing really happens'. It is the opposite that they associate with, and enjoy in the city. In other words, city dwellers typically enjoy living in lively, energetic surroundings which are full of vibrant life, and where they are almost spoilt for choice about what to do and how to use their time. But the city is not only valued for offering exciting individual experiences. Many of our interviewees mentioned that they want to enjoy good communal relationships with their neighbours. As Carl (68) from Amsterdam says after praising the city's social life:

> What's great about this city is the chance to have a good coffee and look at the rain through the window, to read the newspaper, to talk to strangers like we are doing now.

Strikingly, from our interviews we learn that people value living close to lively areas of the city with much going on, and the ability to make choices about how to spend their time and energy, including the ability to take respite from the bustle to visit very peaceful locations. Yet often interviewees did not want to live in what they regarded as noisy, crowded streets. Rather they wanted to live with a sense of community, on good terms with, and known by, their neighbours. There is almost a paradox here. People want to recreate a type of communal village life, where parents and children picnic together in the local park on a sunny afternoon, but with easy access both to the vibrancy of the city centre and the calm of the countryside (or at least a large park), and to feel welcome and at home in all these disparate places. Looking at this in more depth, this is not a paradox at all. City-zens appreciate lively surroundings, but while they want the freedom to make use of them or not, depending on what suits them best at the time, they don't want to feel either dominated by the chaos, or as if they are only visitors to the exciting parts of town. They behave in ways that are in tune with 'the intrinsic logic' of their city (Löw 2012), or its 'spirit' (Bell and de-Shalit 2011), and they typically also want to understand the history of their city, maybe even of their neighbourhood, and their own relation to it, whether it goes back generations or is new. Through their choices, interests, and relations with other people, with buildings, parks, facilities, and history, their life in the city takes on layers of meaning for them. Having a sense of community with others who work or use the facilities of the places one goes to, including the livelier parts of the city, helps create and consolidate this sense of meaningful urban life.

Now, remember that our point is not only that this is an ideal experience of the city, but that it should be equally accessible to all. How can the need for a sense of meaningful urban life for all be translated to, and implemented in, policy? Does it make sense to demand it from the city's authorities? Any such policies would surely raise paternalistic, if not authoritarian, concerns. The role of the city is not so much to provide meaningful lives directly, but rather to provide many foundational elements as a platform for people together to develop the activities and relationships that give shape to both the neighbourhood and to individual life. There will be many aspects to this, from giving licences to restaurants and bars and encouraging independent small businesses, to keeping the streets safe and secure, to providing community centres, night classes, good transport into the late evening, and libraries. Consider how Yap (67) born and raised in Amsterdam, where his family has been living since the eighteenth century, describes his beloved city:

> What I find most precious about Amsterdam is its history, both regarding my own history, and the one of the city. Amsterdam offers everything one might need: culture, social life, accessibility. Yesterday we went to a small square near our house, where there was live music, and a nice cafe nearby served food, and all right outside of our home.

What Yap described is not merely a story about 'how we had fun'. It is the feeling that when you can do, and do, all this, you feel attachment, you feel part of the city in a strong sense, and you feel that your attachment to the urban space in which you live, your city, has a special meaning. Inclusivity was mentioned time and again in the context of policies vis-à-vis gay people, immigrants, the elderly and the young, and how working-class residents could or could not access services in the city. But inclusivity is also about how a city-zen feels in everyday activities and in what many of us take for granted, such as sitting in a cafe, meeting friends in the pub, or having an opportunity to express your culture and customs. Notice, also, that all this is not only about self interest. It is also about how we as a community live, what our city stands for, how we treat immigrants, or what kind of help we offer the less fortunate among us, or whether we respect elderly people. When all this is secure, people feel that their life in the city is meaningful.

Indeed, some cities encourage cultural festivals, or work with communities to provide displays of local history in public libraries. Small, independent, bookshops often try to feature the work of local residents. It is important for many small businesses to employ local youngsters. In July 2022, when we presented parts of this research in Jerusalem, a person in the audience said he buys books only at a local second-hand bookstore called Re-book, because they employ local youngsters with special needs:

> I know it is slightly more expensive than buying second-hand books on the Internet . . . and with the number of books that I buy you might think that I am irrational, but it's about making buying a book much more meaningful for me. You see, even buying a book becomes part of what it is to be a city-zen of Jerusalem.

Another person told a similar story about a second-hand clothes shop. Many Jerusalemites come to these shops to buy their books or clothes because they feel that this is part of what it is to be a Jerusalemite; that you push aside financial considerations and live your life in a more meaningful way. As one Jerusalemite once said, referring to the hedonistic way of life in Tel Aviv, 'in Tel Aviv they know how to live; in Jerusalem we know why we live' (Bell and de-Shalit 2011, 21). Other examples include volunteering within the community, free courses on local urban history, newcomer clubs, street signs in different languages to cater for communities of immigrants and to show respect, and volunteering at the local community garden, a shared space where residents of the neighbourhood can together grow vegetables, flowers, and sometimes picnic together.

City-zens often refer to political activism in the city as part of what constitutes the meaning of urban life. Valentina from Rio de Janeiro describes how significant music and bars are for her urban life, but makes sure we also hear about her political activities:

> I enjoy the outdoor life of the city, where we get to experience the Brazilian culture, including bars, music (especially samba and funk) and dance performances. I also like to be involved in Rio de Janeiro's social movements. I'm affiliated to the Brazilian Communist Party, and I've participated in many political protests (either as an ordinary protester or as a member of the marching band) to fight for better health and education systems.

Indeed, we find that leading a meaningful urban life is clearly the motivation for action when people choose to participate in local politics. For scholars studying participation on the state level, this might appear odd. Being active and participating in politics is often associated with trying to achieve something, securing your rights, or achieving a benefit for you or your community, whether it is a material interest or persuading the legislators to regulate something which you believe is just, or will work for your community's advantage. In other words, on the state level participation is usually instrumental. However, evidence from perhaps the current most innovative form of participation in urban politics—Participatory Budgeting—shows that the main point is to participate, almost regardless of the consequences. Those who participate tell how they enjoy the very idea that local politics becomes meaningful, it is 'ours' so to speak, bottom-up, and not imposed by some alienated council or mayor.

Participatory Budgeting is worth exploring in a little more detail. It is a process whereby the municipality reserves a certain portion of its budget for spending on priorities determined by the community and issues a call for ideas. The sums can be very modest, such as several thousand dollars for each winning project, which can finance, for example, the formation of a local choir, through to much larger sums which can help renovate schools, such as $35 million in NYC (CEC NYC n.d.) and as much as 70 million euro in Barcelona and 100 million euro in Paris.

Participatory Budgeting processes take various forms. Some are mostly online, as is the case in Madrid (Decide_Madrid n.d.) whereas some use a local church or school to provide a setting for face-to-face gatherings, such as the participatory budgeting process in Glasgow, with the aim of tackling child poverty and child obesity, and where part of the goal was to create forums for discussions between residents, including—this was clearly a goal—'ensuring the needs and aspirations of people with disabilities were equitably represented' (Harkins 2019, 11). In the Glasgow example sums of money allocated were not terribly high, around £35,000 for all winning projects together, so each project could hope to receive just a few thousand pounds. Accordingly, the main idea was not primarily instrumental in the sense of getting something from the municipality. It was, rather, to enable groups that are often more vulnerable and disadvantaged to raise their voice, to participate, to feel that they belong, that both politics and their urban life, their life in Glasgow, are also meaningful, and perhaps build on this particular experience to generate new activities. Glasgow's *Centre for Population Health* studied and analysed the process in the city. Their conclusion is illuminating, because their report does not even mention the consequences, in terms of how the money allocated was spent. Instead, they describe how meaningful and democratic the process was:

> In the broadest terms our work has found that PB has the potential to energise and empower communities and to transform and enrich the relationships between citizens, community groups, community anchor organisations and all levels of government and public service. PB can be an effective means of deepening democratic processes and enhancing local participation. PB can illuminate community aspirations and priorities and provide clear direction as to the ways in which service delivery can be improved and potentially co-produced.
>
> . . .
>
> Our work highlights that, like all democratic processes PB is imperfect. However, when it works well PB can be a process of significant learning and collaborative development for those involved.
>
> **(Glasgow Centre for Population Health 2022)**

Such activities and facilities help provide a sense of meaning both for those from the cultural groups who are celebrated in public activities, and for others who can enjoy and learn from something new. There is, however, a potential tension here. Interviewees both valued living in a community with people like themselves, but also often valued diversity (our next core value). How this tension can be managed successfully is perhaps one of the major problems of our age, and to this we turn next.

5.2.3. Core Value 3: Diversity and Social Mixing

Diversity and social mixing is a third core value and a natural follow-up from the second, incorporating a sense of social fluidity. It includes access to lively and diverse shops, restaurants, bars, and entertainments, the presence of a variety of types of people on the street; feeling the city is inclusive with respect to gender, age, and race and so on. As one interviewee said, the egalitarian city should regard multiculturalism as an asset, as an opportunity, rather than as a challenge. Many of our interviewees showed their appreciation of the idea of 'cities with lots of flavours'. Indeed, flavours are a good metaphor: in Amsterdam the Milkshake Festival is celebrated, a multi-genre dance party, where

> life is just a party thanks to the great diversity of skin colours, religions, sexual preferences and male and female forms (. . .) with a clear message where respect, freedom, love, tolerance and fun are of paramount importance.
>
> **(Discover Amsterdam 2023)**

Such practices are already thought to be common in many cities. We think of them as cases whereby the city does not merely tolerate all city dwellers regardless of their race, ethnicity, gender, sex orientation, and so on, but actually takes pride in all of its city-zens and celebrates their presence in the city. In sum, in a city of equals each city dweller is proud of their city, and their city is proud of them.

So a city of equals is a city of plurality and diversity. This is almost another paradox, because often people—especially critics of equality—associate equality with sameness and therefore with conformity. But that is not our aspiration at all, and we did not find among our interviewees any desire for greater homogeneity in the city. Rather, a city of equals is one in which, ideally, everyone, whatever their attributes and tastes, can feel in place, at home. Hence making space for diversity in the city is egalitarian. Compare, for example, a city where transport is difficult and time-consuming, and workplaces are not accessible, and so people with disabilities tend to stay at home, with another city that has taken thoughtful measures to make the environment usable and comfortable for everyone, whatever their disability status, and so city-zens with disabilities make use of, and pass through, the city on similar terms to all other city users. The presence of such diversity is surely a feature of a city of equals.

To push the sense of paradox further, an egalitarian city could include greater economic inequality than its surrounding region, if it can make all people, rich or poor, feel at home. But it is critical that there is mixing and experience, not mere presence, of diversity. There is little more expressive of inequality than extreme residential segregation, especially by skin colour. This is not to say that there is an objection to different neighbourhoods having very distinctive characters. We are not aware, for example, of any significant opposition to the existence of Chinatowns in many major cities, or to enclaves such as the Portuguese-speaking community in South London (South Lambeth Rd) who live in a tight, self-selecting, community. As discussed in Chapter 2, we also need to keep in mind the debate especially in the United States between thinkers such as Elizabeth Anderson (2010) who argues for 'The Imperative of Integration', while others, such as Ronald Sundstrom and Tommie Shelby respond that liberals who insist that neighbourhoods must be mixed misinterpret what African Americans and other minoritized groups currently want in American cities, as many want to feel that their neighbourhoods reflect their own values and culture, whereas moving to mixed neighbourhoods would always place them in a minority. The idea of being able to choose where you live and whom you live with seems very appealing. But at the same time, it seems highly problematic from the point of view of equality when minorities are pushed into inner-city ghettos, or out to poorly connected outer suburbs, not only because these areas will lack facilities available in wealthier areas, but also it is likely to lead to segregation not just in housing, but in work and leisure too, leading to far-reduced social mixing.

Young is also critical of integration per se (2002, 203–4). We have discussed some of her ideas in Chapters 2 and 3, but here we wish to elaborate, as we find her work very relevant to the core value of diversity and social mixing. Referring to the urban setting she argues that integration as mixing is not necessarily the best model for creating inclusive democracies. Group-differentiated residential and associational clustering is not necessarily bad in itself, she claims, provided that it has arisen from legitimate desires to form and maintain affinity grouping, and any spatial group differentiation is voluntary, fluid, without clear borders, and with many overlapping, unmarked, and hybrid places. Young argues that it is reasonable to assume that many residential patterns result at least partly from a preference that members of these groups have for living near those with whom they feel affinity. As mentioned in Chapter 2, in Peter Marcuse's (1997) terms, these neighbourhoods may simply be ethnic enclaves rather than ghettos resulting from exclusion by the white majority.[7] While it is hard to object to people choosing where to live if they have the means to do so, one worrying trend is for the privileged to take dramatic steps to isolate themselves from others. In her evocatively

[7] Marcuse develops conceptual distinctions among these three types of residential patterns. An enclave is a clustering of persons according to affinity groups, whereas a ghetto is the exclusion and confinement of a subordinate group by a dominant group. A citadel is an exclusive community of class and race privilege, from which others are restricted access. An enclave is a positive and empowering social structure, according to Marcuse, whereas a ghetto perpetuates disadvantage. Many spaces of racial or ethnic concentration share characteristics of each.

named study *City of Walls: Crime, Segregation, and Citizenship in São Paulo*, urban anthropologist Teresa Caldeira has used the term 'fortified enclaves' (in Marcuse's terms this would be 'fortified citadels') to describe gated communities with guarded entrances and private security, which have become common in some of the most divided cities in the world, and are a powerful symbol of inequality on all levels (Caldeira 2000).

Realizing that residential segregation comes in different forms, Young writes:

> If residential concentrations simply reflect a preference for living near certain kinds of people, then their existence should not present a problem. But how do we tell the difference between residential segregation and residential clustering in these multicultural cities?
>
> (Young 2000, 203)

Now, Young's main concern is clearly racial segregation, although in Chapter 3 we noted that our interviewees were also concerned with other many other forms of discrimination. We are also concerned about patterns that create ghettos as the effect of explicit or implicit collusion. So, in the following points, while remaining very heavily dependent on Young's analysis, we take the liberty of slightly modifying Young's criteria distinguishing between voluntary residential concentrations, which reflect a preference for living near certain kinds of people, and segregation which is exclusionary:

1. If studies show that migrants, elderly people, working-class people, or others marked as racially or ethnically different experience housing discrimination in majority neighbourhoods, then this means that many members of these groups are confined in their housing options to racially, age-oriented, or class-oriented, concentrated neighbourhoods.
2. If residents of the city 'know' where racial and ethnic minorities, or immigrants, or the less-affluent, are said to be living, and if these neighbourhoods carry associations of danger or boundedness to city residents, then those living in them are likely to suffer stigma that affects other opportunities.
3. If members of the majority cultural group are moving out of neighbourhoods associated with racialized groups or with immigrants there is probably a segregating process.
4. If both public and private resource and property owners fail to invest in the racially concentrated neighbourhoods, or in neighbourhoods where the less-affluent reside, and the latter decline in quality, there is probably a segregation process.

5. If the neighbourhoods in which racialized groups, immigrants, or poorer people cluster have disadvantages compared to others, such as having weak transportation access, low-quality housing for the price, location near unpleasant or polluting industrial facilities, and so on, then the cluster is partly a matter of privilege.
6. To the extent that discriminatory attitudes and behaviour force or induce members of racial or ethnic minorities to live in certain neighbourhoods when they might otherwise seek housing elsewhere, or if people from minority cultures find themselves in such a situation, they live in segregated conditions.
7. Even more importantly if their housing conditions, neighbourhood location, and general quality of residential life are inferior, then their segregation contributes to conditions of structural inequality.

Young encourages a shift in the focus of inclusiveness from spatial integration alone to economic participation in the labour market and political participation in the polity of the city. This reinforces our point from Chapter 2 that space does not capture the entire picture when it comes to equality in the city. Thus, she argues inclusiveness means integration to labour markets and to political positions, and not only at the neighbourhood level, and this is obviously true with regard to gender in the city as well as race. It is also highly important to see that there is no hidden, informal exclusion from political positions for women and minoritized groups despite legislation and regulations that prohibits it. Not only, typically, are there far fewer women than men as councillors (Fawcett 2022), but in some cities female council members testify that while formally they could join committees and debates, in practice they often suffer from the chauvinistic attitudes of many of the male members in those committees (Drage 2001). No doubt similar comments could be made about race and disability status.

We have just argued that a city of equals should not be defined in terms of housing settlements and space alone, and yet, as should already be apparent there is such a difficult balance to be struck with regard to patterns of residence. So we must concede that it will often be a critical factor, and feel the need to return to the topic, especially of apparently voluntary self-segregation, and look at some more examples from our interviews. While a segregated city seems almost a paradigm of inequality the fact that minorities often want to live together came out starkly in interviews with many ultra-Orthodox Jews in Jerusalem. A significant proportion of our interviewees often wished to live in a segregated neighbourhood or town, and this is, at least in part, due to their rabbis and leaders teaching them that exposure to other cultures is a threat to their purity, religious strength, and spirituality. A cynical sociologist might claim that this is yet another way the leaders control their followers and make them obey the leaders without a second thought. False consciousness would be another dismissive way to interpret what we were told. A more charitable interpretation would acknowledge the wish to retain a culture and system of beliefs in the context of a modern, liberal, or at least partly liberal country, which provides a constant challenge to their world view if not held at arm's length.

An unpublished seminar paper by Adi Ben-Dahan discusses the pros and cons of voluntary segregation for minorities, based on a series of interviews with twenty ultra-Orthodox men and women in various neighbourhoods of Jerusalem, both mixed (ultra-Orthodox, modern religious, traditional, and secular living in the same area) and ultra-Orthodox only.[8] Based on these interviews she argues that some ultra-Orthodox people are very happy to remain within a monistic neighbourhood whereas others see the advantages in living in a pluralistic neighbourhood. Although this is our speculation, it would not be surprising if what is said here would also be typical of other groups around the world who wish to live their lives based on a very strong religious faith. Some had no doubt that living in a self-segregated neighbourhood was to be favoured. Yedidah, aged 20, male, says:

> Of course a separate neighbourhood is preferable; no question at all. I don't wish to tackle what I don't need to tackle and see. Each according to his lifestyle, the more separate the merrier. It's not because I have a problem with secular people, but when you live in mixed neighbourhoods it only creates problems.

Hedva, female 29, says that she was looking to live with very similar people, because 'one needs to feel one belongs' (she means in cultural and spatial terms). Plus, living with others might be a threat to your way of life. Jonatan, aged 18, explains that

> there are things one does not want to see, for example breaking the law of Sabbath, dressing in a non-modest way. What is the threat? This can harm your spirituality, this is confusing.

But even those who see the advantages of living in a segregated neighbourhood can also experience it as oppressive: Netanel, male, aged 20, says:

[8] We should add that among the ultra-Orthodox there are many different castes and schools of thought, and tradition, and some extreme ultra-Orthodox wish to reside only with those who obey the same rabbi.

> The advantage of an ultra-Orthodox-only neighbourhood is that you live with people whose behaviour is exactly like yours. But the drawback is that sometimes this is annoying—everybody knows everything about everybody else. You just have to behave the way they want you to.

For these reasons many of those who prefer to live in a mixed neighbourhood justify it in terms of the liberty it gives them. Jacob, aged 57, says that 'in such a neighbourhood you owe nothing to nobody', whereas in a monistic neighbourhood you must conform to the norms and what you do is everyone's business, and they are all watching you all the time to make sure. But Jacob immediately adds that if the neighbourhood is too pluralistic then he would feel uneasy: 'People wish to feel comfortable; and they feel so when their neighbours' life style is not too different.' Shoshi, aged 17, also agrees that when people are all the same you must constantly take into account they are watching you, whereas in her mixed neighbourhood, where she prefers to live, she does not need to please anybody; she does what she pleases. Annette, female 48, explains that she and her husband always looked for a mixed community so that they could live with open-minded people.

Similiarly, Ruth, aged 23, is less worried about what happens to her purity and spirituality. She claims that when she lives in a mixed neighbourhood she learns and adopts a form of tolerance: 'People just seem fine and so I learn to be less judgmental of them.' Rivka, 47, says 'only by living in a mixed neighbourhood did my children learn to know the other and to appreciate the way we live without disrespecting the way others live'. For the same reason Eliezer, 27 male, prefers a mixed neighbourhood, and he adds, perhaps prejudicially, that the monistic ghettos are always dirty and never well kept.

For ourselves we do not want to be judgmental about those who wish to remain within their particularistic community, although admittedly this way of life is not our choice. However, we do wish to assert that a city in which sub-communities are totally sealed off from each other is at odds with the idea of a city of equals as we see it, and in particular the idea of diversity and social mixing, the core value we are currently discussing. But at the same time, we accept that social mixing first can, under the right circumstances, be compatible with the existence of residential enclaves. And this is because, in our advocacy of social mixing, we do not take the position that social mixing needs to be central to the life of everyone in the city. That itself would be to be too directive and intrusive in giving directions to how people should lead their lives. Rather, in our view in a city of equals people have a sense of social fluidity, which means that they should not feel trapped in their locale, but have the ability to move through the city, and, correspondingly, to welcome others to their neighbourhood. There is a delicate balance between preserving traditions and embracing change, and different people will have different taste and judgement here. Of course, it may be impossible to satisfy all.

The discussion above refers to a widely shared dilemma in many cities: how can we cater for the preferences of those who wish to live with similar people, and yet refrain from preventing others who wish to live in the same area from living there? In Chapter 4 we discussed the issues of gentrification and urban renewal; they also become critically important in the context of the current discussion, in terms of what types of renewal can treat everybody's interests and preferences equally. This is not necessarily about the city authorities because the process is not always directed by the city municipality. It is sometimes encouraged by the state (Shmaryahu-Yeshurun 2022), and more often by the market. Nevertheless, even when the market is the main determining force there remains the question of how individuals, NGOs, civil society behave and act.

Now, obviously, there is a difference between people claiming that they want to live in a neighbourhood with people who share with them a culture or a religious belief, and people who claim that they wish to live in a neighbourhood where everybody is, say, white. The distinction is between cases in which the minoritized group expresses such a wish, as in the cases discussed by Sundstrom and Shelby, and cases in which the majority deceptively uses the claim 'I want to protect my culture' in order to cover up racial discrimination. It seems to us that, for example, a city of equals can tolerate, even embrace, cases in which immigrants from Arab countries to a large city claim that they wish to live side by side with people like them, to have their mosques, etc.; but a city of equals cannot accept cases in which white Christian dwellers of the same city put forward a claim according they wish to live only with Christians in their neighbourhood. This may seem very hard to explain. But there seems to us an important distinction between a situation where a minority culture is fragile and needs a certain degree of protection to survive and cases where members of the majority hide behind implausible claims that their culture is under threat as a rationalization for racial discrimination and protecting their own group's privileged access to superior resources.

An even more difficult case is that of a minority that forms an enclave with people like themselves, but then discriminates against a minority within this minority, most notably on the basis of gender or sexual orientation. This is a serious problem concerning the limits to liberal toleration, and we accept it shows that there can be tensions within the idea of equality. We do not believe that there is a simple answer to such dilemmas, but recognizing and respecting the standing and voice of all within a community will at least be a start towards breaking the stranglehold of long-standing prejudice.

Our overarching question, however, is how to achieve a sense of fairness and equality for all, while respecting traditions of different groups, often identified with specific locations in the city, and drawing on different ethnic and religious heritage. A case study in how this could be managed is taken from what was reported to us about school lunches in Paris, where 80 per cent of children eat lunch at school which, whether or not it is accurate, is a helpful model to reflect upon. The regulations, so we understand, lay down rules about the nutritional value of the food, setting out the expected calorie content, and that it should be fresh, prepared the same day, and based on local ingredients. However, each quarter has the autonomy to supply different food, sensitive to local preferences including those of groups of minority ethnicity.

As we have noted, generally speaking multiculturalism was widely appealing to our interviewees. Linn, 25 Belgian by origin, has been living in London for some years. When reflecting about what equality in the city means she comes forward with a single sentence which we are happy to embrace. 'The multiculturalism . . . So that's what I like about this. There's no monolithic culture here, so to speak.' A city comes closer to the idea of a city of equals in this dimension the more difficult it is to characterize the monolithic and dominant culture. This is not to endorse a dull homogeneity. We do not ask for every city to resemble each other in the way, perhaps, their international airports or up-market shopping malls do. We have emphasized that each city will have its own character, but we argue that this needs to be a distinctive blend, not the domination of a single flavour. Fatima, a Muslim Berliner, aged 24, previously lived in Hamburg and Mannheim says:

> Well, what's really better here [in Berlin], in comparison with many other places, is that we've got many people from many different places here, which means that it's very diverse.

In discussing social mixing we have particularly emphasized people from different backgrounds meeting together. But we should not overlook the importance of people being able to mix with others more generally. Loneliness, going beyond the welcome 'anonymity' we have discussed, especially for elderly people, is becoming recognized as a major social problem, both in itself, and in its consequences for health and, of course, for community. It is often assumed that it is a particular problem in cities, which it may be, but cities also have the resources to do much about it. One way of attempting to solve the problem could be through redistribution of income, so that more people can afford to go out to pubs, restaurants, the theatre, and meet more people, although it may have to be very significant to have more than a minimal effect, and indeed these are all places where people, while being with others, rarely interact with strangers on more than a very superficial level. It could be more effective for the local municipality to try to subsidize clubs or build dedicated places—cafes, classes, etc.—for elderly people, as well as for lonely people at all other life stages, to meet others. Or put benches on the pavements so that anyone can take a walk knowing that there's a place to rest and watch the world go by and at least feel part of a wider social world. And of course such facilities are more likely to be used by older people. Another possible policy is to allocate special time for elderly people to use public facilities such as swimming pools, because some may be embarrassed to swim next to young swimmers, feeling that they 'stand (or swim) in their way'. Interestingly the alienated city can offer more opportunities in this respect even than some deliberately communal forms of living. An unpublished interview study compared the experience of elderly people in kibbutzim (communal, collective farms) and moshavim (semi-collective farms) and the experience of people a similar age in the cities. Contrary to expectations elderly people in the cities felt much less lonely, not only because there is more chance that other family members live close by and can visit, but because they can walk outside in the busy streets, on pavements with benches, where they can rest, and engage in unplanned conversations, whereas in the moshav and kibbutz there were no places to stop. Elderly people in the city went out for a walk more often even than those a little younger than them in the village.

But to return to the issue of social mixing between people of different backgrounds, as we have noted, a balance must be struck between preserving traditions and embracing change, and it will be very hard to solve this problem in a way that satisfies everyone. But it is vital that no one has good reason to feel excluded. Which leads us to the fourth core value.

5.2.4. Core Value 4: Non-deferential Inclusion

The term 'non-deferential inclusion' may be slightly unusual. What we mean is that all people should be able to receive the same type of treatment without some having to show particular gratitude or deference on receipt, or go through processes, without good reason, that others are spared. All individuals are to be included in social, political, and economic life on equal terms with others, rather than grudgingly or in ways that humiliate them, such as making it clear, implicitly or explicitly, that they are, or could have been, in some sense, outsiders or second-class citizens, who should be grateful (cf. Wolff 1998). The idea is that inclusion is a right, not a mere privilege, but also such a deeply held and respected right that no one need even mention it. When waiting for a bus, others don't push past you, ignoring that you are even there. When waiting at a bar, you are served when it's your turn, and not after all the wealthy, white men have been served first. When claiming welfare benefits you are not made to feel that you have lesser claim than others, and need to wait and show special gratitude, because you are an immigrant. When you apply for affordable housing or a kindergarten for your child you are given consideration when you reach the top of the list, no matter your religion, your gender, your sexual orientation, the colour of your skin, or who your ancestors are. When the city authorities plant trees to create shade, which is becoming more and more important due to climate change, you don't have to fight for the municipality to plant trees in your neighbourhood as well, even if you pay less local tax, because you earn less compared to your boss, who lives in the nearby affluent neighbourhood; and when the municipality decides where to locate a garbage processing plant, they do not consider only those neighbourhoods where the less-powerful, less-connected, and poorer city dwellers live; and so on. It is easy to see the intrinsic connection with a sense of belonging, but also a recognition of the belonging of others.

Another way of putting this is that we can distinguish two 'levels' of inclusion. The first is simply to have your basic rights recognized and satisfied. The second is for those who you are dealing with to recognize that the rights are the rights of a human being with a whole set of needs and interests connected to each right, and there are ways in which rights can be fulfilled in a minimal way that ignore these surrounding needs and interests. Only if these other factors are taken into account has one truly been treated with respect.

The idea, then, of non-deferential inclusion is the feeling of being part of what constitutes the city, of what makes it what it is;[9] being treated fairly and with respect by the city and other city-zens; people having access to the facilities and resources of their city by way of automatically assumed

[9] In some cities the more affluent and advantaged feel embarrassed to associate themselves with the city, which is thought to be poor or deteriorating. This was the case in cities like Detroit that went through an economic and demographic decline in the 2000s and early 2010s. In Brighton we heard that residents of the more affluent nearby suburbs are said to reply 'Hove, actually' when they are asked whether they are from Brighton ('Brighton and Hove' is the formal name of the city), as the Brighton part has a high percentage of deprivation and the city has the third largest number of homeless people in Britain (Berry 2021) and is thought to be a lower-middle-class or working-class city. Thus, in that respect, the more people who come from the more wealthy neighbourhoods associate themselves with the city as a whole and answer that they are from the city, the more the city is a city of equals.

entitlement, rather than having to throw themselves on the mercy or discretion of gatekeepers. An important parameter is whether officials listen to everybody.

What does feeling part of what constitutes the city mean? It means that one feels, for example Shanghainese, or Istanbulie, or a Porteño, or a Londoner, and not merely a person who happens to reside in Shanghai, or Istanbul, or Buenos Aires, or London. It means that your dialect or accent does not make you a stranger or even unwanted. It means not having to justify your customs, and so on, at least as long as you don't harm others. It also means that you can be who you are in public without being ashamed or embarrassed, such as same-sex couples walking hand in hand in the street, without others staring or pointing or showing disapproval.

Non-deferential inclusion implies that people should feel welcome anywhere they go. Earlier we reported the remark of Amalia, a religious woman in her twenties, who told us that Jerusalem is 'the most egalitarian city' she knows of because 'as a religious woman I can go to a bar here and feel free'. One can sense the diversity, she says, and yet it is not an obstacle. (We might wonder whether this is true of Arab city-zens of Jerusalem, or of secular female Jerusalemites who attend a restaurant in the ultra-Orthodox neighbourhood, but if Amalia's picture is true, then this is a sign of a city coming closer to the idea of a city of equals.)

Inclusiveness also means that the newcomers (immigrants from within the country or from outside the country) and also those previously socially marginalized are included in the job market, cultural events, social initiatives, local politics, social networks, local NGOs, and so on. Inclusion can be a quantitative concept, measured by statistics, such as the numbers or percentages of the newcomers or the previously marginalized and excluded who are now taking part in planning, deliberation, politics, social activities, cultural activities, or whatever parameter we can think of. (And this can be measured either subjectively, that is to say by relying on people's reports, or objectively, by comparing statistics and so on). But we are equally interested in inclusion as a qualitative concept, and in particular the quality of relationships. This again can be found by interviewing people and asking them about their feelings; or by asking what the expectations of newcomers and the hosting community members from each other are, or what do they think should happen.

It is also possible to explore the quality of relationship more objectively by interpreting what we see. In our previous work *Disadvantage* (Wolff and de-Shalit 2007) we identified a number of functionings that have to do with the level of relationships, which most obviously includes the functioning 'affiliation'. However, we also added a number of additional categories to Nussbaum's famous list of functionings, or, rather, in her terms, capabilities, including: (1) understanding the law; (2) being able to communicate, including being able to speak the local language (where the local language is broader than language, it is also about norms, values, etc.); and even (3) doing good to others. All of these are relevant to the idea of non-deferential inclusion. Consider the last. One way of being excluded is by people declining your hospitality, if, for example, people shun a free festival put on by an immigrant group.

Related themes regularly occurred in our interviews, as we reported in Chapter 4, and it is worth revisiting these ideas for the insight they provide into the idea of non-deferential inclusion. Homelessness, and especially street sleeping, is a particular case in point. No city will completely ignore homeless people, or those unable to feed themselves, and there will always be some sort of provision, whether through the city or the voluntary sector, including religious groups. But what level of provision is on offer, and how people are made to feel about using it, can vary tremendously. Do food banks insist on minute examination of your paperwork and send you away as soon as you have your allocation, or are you invited in for a cup of tea and a chat, and made to feel, at least while you are there, a human being? Tina, 49, from Hamburg, complains not about the presence of homeless people, but about the lack of support that they receive from the city, and similar sentiments were expressed many times by our interviewees. City dwellers nowadays accept that homeless people are attracted to their city or that due to rising accommodation prices many become homeless; but they strongly believe that therefore the city should do more to help them, assist them, with food, medical treatment, and shelter.[10] Andrew (32) from Oxford echoes Tina's views about the lack of much-needed policies to assist homeless people. He says:

> I don't blame the homeless. It's not these people's fault at all. They are victims of the system. (. . .) The council doesn't do anything. They let them do whatever they want, as if they didn't care. There should be more shelters. But this system, in a way it seems as if people have simply accepted that [the poor's] children will also be poor.

We have also already noted the comments of Jason, a young Black German in Hamburg, who remarked that while he believed that the city is 'truly multicultural, especially compared to other cities I have been to' nevertheless on

[10] For an interesting discussion of what a city's obligations towards the homeless are, see van Leeuwen (2018).

his first day at work at a construction company 'my colleagues met me and the first thing they asked me was "You're not a Muslim, are you?"' It is interesting that he was not denied the job, so in that sense was included, yet his workmates were not instinctively prepared to accept him on equal terms. Now, they may defend themselves by claiming they were just having a bit of fun and making a joke, but what counts as an acceptable joke sends a message; that Jason, unlike his workmates of European descent, has to prove himself before he is included. This, therefore, is an example for us of deferential inclusion—at least at first, he had to acquiesce in treatment that others, most likely, did not need to suffer.

There are also less noticed, but no less fateful, patterns of exclusion. We were struck, when considering measures of relative poverty, that one item that comes near the top of the list on some studies is 'being able to visit friends and family in hospital'. We speculate that most academics reading these words would be very surprised, as they have the financial means and flexible schedules to be able to visit their friends and families, and simply take it for granted that this is not so much a privilege as a matter of ordinary life. But those who work inflexible shifts, or lack a car or money for travel, or have extensive caring relationships with no support, are in a different position. This has been exacerbated in recent decades by changing patterns of hospital provision in some countries, with small, local, general hospitals being replaced with large, specialist hospitals, sometimes on the edge of the city or even in a different city within the same medical authority, or, in the most extreme cases for very specialized care, on the other side of the province or country. There are, of course, very good medical reasons for such specialization and consolidation, but the effect on poorer residents and those with less-flexible lifestyles can be highly problematic. Suppose a family member points out the difficulty of visiting and is told by the hospital or their boss 'you should be grateful that your family member is receiving the right care' rather than being helped to find ways to visit. We would call this almost a paradigm example of a failure of inclusion without (the expectation of) deference: as we said above, a matter of someone's basic rights being met, without full recognition that they are the rights of a human being with other, connected, needs.

This is also an example of how not being fully equal in one way can be associated with, or even cause, inequality in other respects, in this case how inequality in working and living conditions can cause inequality in access to friends and family in hospital. In *Disadvantage* (Wolff and de-Shalit 2007) we refer to this as clustering of disadvantages, and similar comments were made in our interviews. Tid (30) from London explains how sometimes disadvantages cluster, creating a compound which needs to be analysed in its own terms (what is termed today 'intersectionality'):

> OK, so (. . .) in East London, the thing is, you had all this [being not only working class but also often unemployed] and then on top of this, it's racially charged. And also, the rich here, they live exactly on the other side of the street. This racial segregation is there because poverty and areas of poverty largely get inhabited by, you know, people of colour who are migrants, or their descendants, involuntary migrants, many Windrush people really.

Now, we need to address the question of whose responsibility it is to send the message that everyone is equal and included: the city's authorities or the city dwellers? J. S. Mill's answer, we have already noted, is that informal social norms have the potential to harm dignity and liberty no less, perhaps even more, than formal regulations. Jason's story about being asked whether he is a Muslim is about his co-workers. Even if the municipality, or more likely the state, produced laws forbidding racist speech, it won't be able to stop people from thinking this way. But it might be able to make them think twice before they express prejudicial thoughts. This has three advantages. First, expression of discriminatory thoughts is likely to become less common. Second, third parties will be less exposed to discriminatory attitudes and therefore less likely to legitimize them, which might have a positive effect on these other people as well, in the sense that such thoughts may fade from their minds. Third, if people are exposed to such attitudes, they will know that they are entitled to complain, either directly to the assaulting person or body or to the city's authorities or the police.

Sometimes more privileged city dwellers are unaware of, or not sensitive to, such cases of deferential inclusion. This is very clearly explained by Klaus, from Hamburg, who says that the poorer in Hamburg, where he lives, who lack education, feel they will be looked down on if they don't dress up, and so despite not having enough money they spend a lot on things you would assume they don't really need:

> I can afford being looked at strangely and with bewilderment (if I am not dressed up)—but they can't. They feel downgraded. No one needs to starve here, but people feel left out.

Sometimes these are more than thoughts. Alba, aged 38, a Berliner who came from Africa, tells us this story (also quoted in part above):

> I understand it to be prejudices. (. . .) Africans and Arabs are treated differently. Africans particularly badly. (. . .) I was refused entrance to a club. Everyone got in. The bouncer didn't allow me in. I had no drugs, nothing. This is denial of freedom of association. I told that to the police around there, but they just told me to go home. I felt bitter and humiliated.

He argues prejudice is all too common: 'Young women—when a black man approaches her, she will start holding on to her stuff. It's so humiliating.' Alba claims that Hamburg's bureaucrats are also guilty of instinctive prejudice, which is especially important to our analysis. In a city of equals bureaucrats in particular should make everyone feel that they have an equal right to the services of the city. To our question whether this xenophobic prejudice comes from individuals or whether it is something structural Alba answers:

> Five years ago I applied for housing and had all my papers ready and I didn't get anything. So, I had to live in a *Wohnheim* [shelter or residence hall]. All this is a disenfranchisement of people's life. And when you complain you're considered a troublemaker.

Inclusion of cultural, lifestyle, and gender minorities is important, and yet, as we have noted, it should not come at the cost of ignoring class and economic differences (and vice versa). In Chapter 4 we discussed our interview with Lisa, 43, who lives in Oxford, and we reported that she describes the city as extremely sensitive and egalitarian towards minorities, with regard for extensive gender, and especially LGBTQ+ issues, but she claims this comes at a cost:

> See, gay rights are important and high on the agenda, rightfully, this kind of inclusiveness is, I think, very present. There are many students here and it is great for that. (. . .) But I often feel like many are not fully aware of the fact that their rooms are being cleaned by someone, and not aware of their backgrounds or situation. At my workplace, for instance, I just learned that the cleaners there, who work for an outside contractor, they don't get sick leave. I definitely think that many people here, young people, are very aware of the gender and sexuality issue—but it seems to me that there are many other problematic things that they are not so much aware of. Oxford as a city (. . .) is trying to be very inclusive . . . and promoting equality . . . and people are aware, but there is awareness of certain types or directions in particular. I am talking about social and economic inequalities.

Many of our interviewees had in mind quality of relationships when discussing inclusion. When asked about what she would want to be improved in

Berlin, Kathi, a 33-year-old Berliner, says: 'Well, I do think that people have become less and less friendly, and that the number of people here has grown and grown.' This complaint came out time and again in Berlin, and it seems that it matters to city dwellers a lot. She continues:

> I can tell you something (. . .) No 'good morning', no 'good day', only being honked at on the streets, people even spit at you, all this vulgar behaviour—isn't this super annoying? No child says 'please' or 'thank you' anymore (. . .) There is no friendly 'togetherness', community.

The opposite picture was revealed in Jerusalem. Many interviewees said that although the city is very multicultural to the extent that opposite and rival cultures live together side by side, this does not harm the friendly atmosphere in the street and the latter compensates for the political tension. Once again, we are not endorsing the empirical accuracy of these testimonies and it very well might be the case that Berliners, for example, tend to be more self-critical than Jerusalemites, or that some Jewish Jerusalemites deceive themselves about how welcome Arab Jerusalemites feel. But this is not the point. What is important to us, and what we derive from these interviews, are the core values for the city of equals. And what is impressive is that in all these cities and among the many interviewees, the same parameters were mentioned.

Last but not least, generating a sense of non-deferential inclusion depends heavily on the way the city authorities plan, and the ideological intention behind the planning. Moreover, it also depends on the extent to which city dwellers actually fulfil these intentions in their everyday implementation of the regulations and policies. To explain, let us consider urban right-of-way regulation. The public right of way in a city refers to areas on or above a public roadway, street, public pavement, alley, etc. Perhaps more interestingly, it can include privately owned spaces adjacent to private properties, such as paved areas around offices and apartment blocks. Where the public right of way applies, these surrounding areas are to be kept available for public use. It is the equivalent of the right to roam in the countryside: private property which often remains private, and yet must be open for people to use or walk through. City planners also often allocate and design areas around schools, or other public buildings to remain open and available for everybody. The idea that everyone can walk through, without feeling that others are doing him or her a favour, is crucial for this feeling of non-deferential inclusion as part of a secure sense of place.

However, it is sometimes the case, in less-egalitarian cities, where property owners have a lot of power and influence, that despite the planning regulations, property owners put obstacles and make it difficult for others to use the place. For example, in Tel Aviv, the legal status of the right of way is very strong. Any constructor or company that gets a permit to build a house should leave some space for public enjoyment of the area; a little garden, park, plaza. And yet many young residents of Tel Aviv who often cannot afford to live in the more modern and expensive buildings often complain that these open spaces are designed so as not to give people the feeling that they are invited; fences, bushes, trees, even gates which are illegal, create an impression that the place is for private use only.

A way to overcome this is to apply placemaking philosophy in planning, which means that the planner sets out not to build an object, such as a house, a plaza, a museum, a shop, or a factory, but rather to make a place, which is a more holistic concept. This is a matter of thinking not only of the building and how it will look, but also about who will live there or nearby, who will walk by or through, how it will be used, for what purpose and by whom, and how will it be maintained, and so on.

Another way to overcome effective privatization of space is by regulation. A city can disallow signs such as 'private, no unauthorized entrance' or 'no trespassing' at the entrance to apartment buildings and public buildings, such as university campuses, libraries, and so on. Of course, the question is complicated regarding the lobbies of apartment buildings. On the one hand, the owners who purchased the apartments or those who rented them in practice paid for the use of the lobby as well as their own apartments. So there seems an obvious sense in which such lobbies are private property. On the other hand, many such buildings have spacious lobbies which are rarely used and could function as gathering places for others who could make use of them, such as groups of students who want to study together, a local NGO that needs a place to meet, and so on. Why not allow them to use these spaces? This is especially pertinent in cities where real estate is very expensive. So, it seems to us that in some cases a city of equals could introduce measures to allow greater use of scarce space. This is most obvious in cases of public buildings which particular groups regularly use (e.g. university buildings, schools, churches) but could benefit others when and if they need to use them. Passing through gardens in the front of private apartment buildings should be equally unproblematic. For other cases, such as spacious lobbies of apartment buildings, there could be particular arrangements, to regulate times and conditions under which they could be used by non-residents. But this would not be because this is private space or private property but because of respect to the privacy and security of tenants and their right to enjoy quiet evenings and nights. Whatever the details of the arrangements, such approaches take into account the plurality of the city dwellers, and their different levels of access to resources and in that sense should sustain non-deferential inclusion, as well advancing all the other values: non-market access to facilities, a sense of a meaningful life, and diversity and social mixing.

5.3. Conclusion

The purpose of this chapter has been to introduce the details of our theory of what makes a city a city of equals. Our general conception, repeated many times, and based on reflection on our interviews, is that a city of equals is a city in which all of its residents feel that they are part of the city's story, and enjoy a secure sense of place. This overall characterization of a city of equals derived from the four core values that we defined in the following way:

1. Accessibility to the city's services is not constituted by the market.
2. A sense of a meaningful life.
3. Diversity and social mixing.
4. Non-deferential inclusion (that is, being included without having to defer).

We finished Chapter 4 by referring to possible tensions between the various themes, forcing cities to make difficult choices. Now that we have expanded further on the themes and core values, we are in a position to explain our approach to the potential tensions and trade-offs between the many different goals that a city of equals should promote. For example, there is tension between permitting boisterous uses of public space and fostering a sense of security. In reply, we would like to make three related points building on each other. First, we accept that this is true, and in reality cities, mayors, and councils will have to prioritize according to their own city's values which may point to different outcomes in different cities, but will often lead to a pragmatic compromise, such as reserved and protected times of use. Second, and developing this first point, unlike material equality, where typically one person or group's gain is another's loss, in relational equality, people are often prepared to make compromises because they benefit in another way, in terms of their sense of belonging to a community with others which is essential to them. Third, and perhaps just a more detailed way of making the last point, these tensions are often diminished and might even vanish when it comes to urban politics, as opposed to national politics. There is a considerable difference between the way people think as citizens of states versus how they think as city-zens of cities. The difference corresponds with the two ways in which politics is perceived. One way is politics as a competition: who gets what and how, which implies that politics is antagonistic; it is about trying to persuade the governing body to prioritize your interests over the interests of others. The other way politics is perceived is as a process by which we find a way to live together despite all our different interests and ideologies, religions, and so on, and the living together despite differences is regarded as highly valuable in itself. It is evident that city dwellers switch from thinking as citizens to thinking as city-zens and that when they do so, they also switch from the first to the second conception of politics. So, many of the tensions between core values can be handled in deliberations, city-zens' forums, and so on. One mechanism which started to spread rapidly and quite successfully is the Participatory Budgeting mechanism which we mentioned.

But we do not want to pretend problems of tension and trade-offs in the city can be made to disappear. City council meetings, not to mention local elections, are often places of intense contestation of values, and local papers overflow with stories of protest at city plans and neighbourhood disputes. The problem of conflict of values will always be a practical, as well as theoretical drawback of any pluralistic theory. Still, we remain faithful to our goal of putting forward a pluralistic theory because it is more realistic than any theory based on a single principle. Thus, we acknowledge that since our theory is pluralistic indeed, there will always be some worries about such tensions. But we do not regard it as a weakness that we have declined to provide detailed solutions for each and every case. When tensions arise, especially when several options all are viable candidates, there is room for democratic intervention, and policies should be decided deliberatively and democratically by each city, as we see in the case of the many models of Participatory Budgeting. This is just one example of how such local democratic processes are possible and when the city-zens of a city find tensions between the core values, they can deliberate about it and find a solution. Nevertheless, we regard the core values as aspirations for any city that takes equality seriously, and in Chapter 6 we set out a wide range of factors that officials should consider if they care about the equality of their city-zens. The thought, then, that there are always many factors that need to be taken into account, but how exactly they should be dealt with will depend on context, history, and local values, among perhaps many other things.

Moreover, we would not by any means claim to have produced a conclusive argument for this picture or any part of it. Rather we have come to this account on the basis of a type of an interpretive triangulation between our own theoretical reflections, including conversations with colleagues and students, the existing scholarly literature, and the many interviews we undertook for this study. We are very open to the idea that there are alternative ways of conceiving of a city of equals, but we feel we have captured the underlying spirit, and expect that any alternative account would at least have a great deal of overlap with what we have done here. The next question, of course, is what can be done with this account now that we have it. And that is the task of the next, and final, chapter.

6
Conclusions and Next Steps

Our project, we have explained, is to come to an understanding of what it is that makes a city a city of equals; a place that appeals to those who consider themselves egalitarians. Combining our own theoretical reflection, reading of the literature, and interviews, we developed an initial central suggestion that a city of equals gives each city-zen a secure sense of place. We then broke this idea down into four critical core values. As explained in Chapter 5, the core values are: non-market access to goods and services; a sense of a meaningful life; diversity and social mixing; and non-deferential inclusion.

The obvious next question is that now we have this theory, what can be done with it? What use could be made of it, and what needs to be done to make it fit for that purpose? It would be understandable to assume that the general idea is to be able to use the theory to judge whether or not any city can be regarded as a city of equals, and to rank cities against each other. But it may also be clear that this isn't quite right, for we have not attempted to set out an ideal of a city of equals and so the judgement that, say, one city has achieved a score of 90 per cent against an ideal whereas another has achieved only 55 per cent, would not be how we envisage using the theory. Furthermore, there is something almost distasteful and contrary to the idea of the egalitarian spirit, as we understand it, to sit in judgement and rank cities as to how much they live up to an ideal. It is like having a ranking of individuals for their modesty. It may be possible, but it rather misses the point of equality, and in any case such rankings carry little conviction. For example, there are annual rankings of 'the world's most liveable city' in which Melbourne, Auckland, Vancouver, Geneva, and Vienna have triumphed in recent years. The headlines these rankings generate can be very useful to get publicity for the organization sponsoring the ranking, and, perhaps, for marketing purposes for the top-listed cities, but few really believe that the ranking settles any important question. Various rankings for cities have been attempted, the purpose of which is not always very clear. Giffinger and Haindlmaier (2010) complain that often public attention of city rankings is mainly concentrated simply on the ranks themselves totally neglecting its meaning as an instrument for strategic planning. Whether they rank cities according to their

City of Equals. Jonathan Wolff and Avner de-Shalit, Oxford University Press. © Jonathan Wolff and Avner de-Shalit (2023).
DOI: 10.1093/oso/9780198894735.003.0006

power (financial, economic, political influence) or comfort (walkability, sustainability, resilience, smartness, beauty, etc.) (Tan et al. 2012) and whether they view cities from outside (e.g. tourists, business people) or inside (e.g. residents, the elderly) these rankings often do not gather data in a scientific manner or use it to any purpose.

But at the same time, it would be very disappointing if we were to stop at this point. We have set out a broad account of a city of equals, with four core values, each of which can be dissected into further themes. To refuse to go any further raises the question of why we have decided to come even this far, especially as we are especially concerned with policy questions. One possible response to the question of why stop at this point draws on a distinction we made in Chapter 1 between questions of definition and questions of measurement. We could simply rest content with the claim that we are only interested in questions of definition and will leave questions of measurement to others. But there are several defects with this answer, the most important of which may well be that it simply isn't true. We are also interested in questions of measurement, and although we cannot settle all such questions here, we owe to ourselves and any readers still with us at this point, to say more about measurement.

And it is clear that many will have some strong intuitions about critically important factors, some of which we may judge to be much more important than others, and some of these factors are not mere measures of inequality but constitute inequalities in their own right. A viciously racist city, for example, will be very far from a city of equals, however well it does in other respects. And, we might say the same thing about a city that makes limited provision for residents with disabilities, or one that fails to offer acceptable means of transport other than the private car, or has any number of other failings. For this reason, it might be tempting to regard different factors as setting threshold conditions for a city of equals, rather than components that feed into a system of measurement. However, even to introduce this discussion is to get somewhat ahead of ourselves. First of all, we do not even yet have an account of the factors under each core value that we wish to focus on, and still less a way of measuring them. These will be minimal requirements before we can operationalize the theory and discuss how the different factors could be weighted or aggregated.

But before deciding what to measure, first we should return the question of what purpose an index is intended to serve. The reason to raise this question, as we argued in *Disadvantage* (2007), is that there is no way of knowing whether an index is fit for purpose until you know its purpose. For some purposes a very simple index will suffice, in others a much more complex one is needed. Of course, once an index is constructed it is possible to put it to unforeseen purposes, but in this case, we want first to be sure what we want before going further. We see two main possibilities; one, the most obvious, which we have already mentioned, is to provide a ranking of the degree to which cities can be regarded as cities of equals; the other is to provide a tool by which a city can monitor its progress (or lack thereof) in the direction of equality, which does not in itself licence comparisons.

Although the general ranking scale seems bold and ambitious, we have already raised a general reservation against it: the apparent inappropriateness of ranking cities for how equal they are. But others may not have such reservations and will wish to push this possibility further. There is no a priori reason why this shouldn't be attempted, but we see at least three problems with the idea that any such index could produce meaningful results.

The first is simply that in international comparisons of cities there is likely to be too much 'noise' in the data, in the sense that the factors under a city's control may often be dominated by national factors. Although in Chapter 1 we argued that cities have considerable power, they nevertheless operate in a context of the national law and national policies, most notably income tax, and the methods by which cities are allowed to raise their own finance. If being higher on the ranking means that the city is more just and its policies are more ethical then there is something unfair if a city that struggles to remain egalitarian in a libertarian state with gross inequality is ranked below a city that is more egalitarian but in a rather egalitarian state. Suppose, for instance, that we do compare cities around the world, and find that in state A, a libertarian and inegalitarian state, there is a city AA that gets a score 0.6, whereas in state B, a social democratic and rather egalitarian state, there is a city BB, which is slightly more egalitarian than AA and gets 0.65. Which city deserves to be named a city of equals? BB that gets a higher score, or AA, that against all odds and within a libertarian legal system gets 0.6? In theoretical terms, what we claim here is that at the end of the day, comparing cities from different states as if the states were not part of the game should not in itself enable us to praise one mayor or local government and condemn the other. At the same time when cities promote many egalitarian policies when the state is conservative and inegalitarian, this is very impressive. One such case is the city of Vienna, which during the Covid19 pandemic issued very egalitarian policies, including generous financial help to small businesses, despite heavy pressure from the central government that opposed the city's policies.

Still, one might argue that the index is not about being fair to cities and mayors but simply to record facts about which city can be regarded more of a city of equals, and hence if a city is equal because of national policy rather than local policy that should make no difference to its ranking. So really this objection amounts to a caveat or warning: if there is an international ranking of cities, performance on that ranking should not be wholly attributed to the consequences of local policy.

A second difficulty may be more important. In seeking an index appropriate for international comparisons, we may be forced to very generic measures that lack nuance. For example, one thing that may have struck the reader is that at several points we have written approvingly of cities that encourage walking and cycling, and give preference to bicycles over other forms of travel. Typically, such cities have an egalitarian character, and it would be tempting to include some measure of walking and cycling in a city equality index. Yet the possibilities for walking and cycling are highly context dependent. Most obviously, the size and geography, such as the flatness or hilliness of terrain, or the mildness of the climate, makes walking and cycling more realistic in some cities rather than others. In some cities, especially in the developing world, people are forced to walk long distances daily simply because of the absence of public transport, but this looks like a way of ignoring the interests of the poor, rather than an enlightened policy. And we should also not forget the importance of social norms. In some cities, with short blocks and a norm of not crossing the road unless one has a green signal to cross, walking even a relatively short distance can be very irritating and time-consuming, whereas in others where such lights are treated as advisory rather than mandatory for pedestrians, while strictly observed by cars, walking across town is much more congenial, provided that the car traffic is not too heavy.

These examples may seem rather trivial, but the problem generalizes. The form of the problem is that any measure has an 'other things being equal' assumption, but when it comes to real life and policy recommendations other things are rarely equal, and we need a more holistic approach than one which a simple index can provide. For example, consider housing affordability. It may seem promising to look at median rents of properties of a certain size or character and compare them to the median net incomes of the lowest quartile of households, for example. But if in one city there is universal health care and in another people have to buy their own insurance policies and/or pay out of pocket then disposable income will be rather different. Of course empirical economists are used to these type of adjustments, but this is only one factor. In some cities it is very common for people to live alone, or even, effectively, to have a household spread over two properties (especially on second marriage) whereas in others for cultural reasons, multiple generations occupy the same housing unit. For another example, some cities such as Toronto or

Miami are extensively multicultural with a very high proportion of residents born outside the country, whereas for cities such as Madrid the proportion, though significant, is considerably smaller. What equality means in such varied circumstances could be rather different, and a generic measure may well have biases.

Once again it may be said that with enough ingenuity these problems can be overcome, but we then come to a different type of issue, one of values. In so far as different cities have their own character it may be that people in different cities simply value things in different ways. For example, street noise in one city may be regarded as a form of pollution, whereas in another it is taken as perfectly normal and barely noticed. A general scale can tell us how a city performs on various indices, but much more work would be needed to convert those indices into an account of what they mean for equality in each individual city.

In one way, though, in the case of our own research we should say the concern about a wide variation in values between different cities looks at first more of a theoretical problem than a practical one, as it was not our experience in the interviews. On the contrary, we were quite surprised how the themes that we have discussed in this book were mentioned in all ten cities and six countries. And yet, there are two points to be made. The first, is that we should recognize the limitations of our study, and admit that because our research is the first to be conducted in that manner, more research has to be done, and it remains to be seen, whether, after such research is repeated in additional cities from all continents, the city dwellers in these cities confirm the themes that we believe are the most relevant to thinking about equality in the city. We take some comfort in the fact that Amartya Sen makes an analogous argument when he writes about which capabilities and functionings should be considered most critically important, suggesting a democratic decision in each country (Sen 1999).

The second point is that even if all themes are shared universally, what we suspect will be different is the relative weight of each of them, such as whether more weight should be given to access to clinics or to playgrounds for children, or to pubs. One could hypothesize that a significant gap in the average age of the city's population could imply different values and different answers to this question. So, our point is that differences at least in the weighting of the core values might be disrespected if we were to recommend a single index.

Thirdly, we would be particularly concerned that an international index would be vulnerable to attempts to game the system; an application of 'Goodhart's law' (Goodhart 1975). Measures would be turned into targets, and policymakers will take steps to improve their performance on the scale, which might not reflect actual underlying performance. To take a crude example, if the scale measures the number of bus journeys, a cynical mayor might be tempted to game the system by coming up with a new way of designating what counts as a journey so that their city rises up the scale but with no one's life improving. This, of course, is a general problem for an issue of measurement, and attempted solutions are possible: auditing changes; having many measures; or making the components of the index highly meaningful and therefore impossible to game. But in the end, is it really worth putting this level of endeavour and resource into constructing, maintaining, and policing an index to make it resistant to gaming if its only point is to rank cities with no further consequences beyond the ranking itself?

Rather, then, the main use we envisage for any such index is not so much to compare cities, but so that a city can understand its own trajectory, looking back over the months and years, and to consider what it needs to do in order to come closer to a city of equals. And just as our core idea is that a city of equals is one that offers each city-zen a secure sense of place, it is also important that the city-zens 'own' any such index as reflecting what they value in the city. This yields two further methodological refinements. First, the index could be customized to each city, at least at the level of detail, if the city wishes to do so, and second the city-zens themselves should be part of that process of customization, greatly involved in deciding what to measure and what to ignore. Having said that, this does not mean we have to start again, as we have made what we regard as significant progress in drawing out the elements that will structure the construction of any such index. It will be broken down into four areas, reflecting our four core values: non-market access to goods; a sense of a meaningful life; diversity and social mixing; and non-deferential inclusion. Beyond this we can break the core values down into themes, which we will do over the next sections, but which specific themes are to be included and how much weight should be given to each of them is, broadly, a matter for each city to decide for itself, although we will offer examples for consideration where appropriate.

Before we get to the discussion of the themes, we should first discuss the types of data that could be used to populate the index. In an ideal world, with limitless time and resources, it would be best to start with a list of factors under each theme, decide on how to measure them, and then commission studies to carry out such measurements. Of course this is unrealistic, and, most likely, the index will have to rely on existing data-sets for the most part. We would like to propose a pragmatic exercise in first finding out what relevant material is available for a given city, and then working out an accommodation between the factors and measures, probably in a modest consultative exercise. Any theory proposed on the basis of this analysis should be scrutinized by the academic community and city dwellers; if the theory is accepted, perhaps after some modification, then it is only if vitally important information is missing that it would be worth considering commissioning new studies, or amending existing forms of data collection, if resources are available. In what follows we will largely, however, disregard this complication, as we will generally be discussing factor measures at a level of abstraction above particular issues of data collection.

A second question is whether we should be looking at 'objective' measurable facts, such as average rents, and the frequency of walking in the city, or 'subjective' factors about how people feel about their experience or the services offered in terms of satisfaction. But the question is partially answered by the previous discussions. First, it will depend on what data is available, and second, on what emerges from the city-specific consultative exercise. But our initial assumption is that both types of data are relevant and important. Below we illustrate this point with the example of how to measure inclusivity and how we need both subjective and objective data to really capture the notion of how inclusive a city is.

Similar comments apply to a third question, concerning whether we should be measuring the provision of services, or their use. For example, should we count the number of public swimming pools in a location, or how often they are used, or, indeed, how many different people will use them in a time period? This relates to a question of how liberal a city wants to be. To rely on Sen again, his argument is that when we think of equality what ultimately matters is people's functionings, but liberals should not measure the functionings: they should not measure whether people do come and read at the library, but rather whether they have the relevant capability: in this case the freedom to read. The policy question will become whether there are well-stocked libraries available to them and no laws or overt social practices that stand in their way. And yet, this is more complicated, especially as social practices can be subtle. To use the example we have just given, if we only count how many swimming pools the city has, rather than whether people do use them, or who uses them, we might overlook the fact that in this city there are obstacles to the use of the swimming pool. Consider the example of city regulations against the use of burkini swimming suits that we raised in Chapter 5. Religious Muslim women in a city that regulates against the use of burkinis might, in some sense, have good access to the swimming pool, but they in fact cannot use them. So once more, what to measure will depend on what data is available, and what the city-consultation prioritizes, but again, on the face of it, there is no a priori reason to exclude any data source, although we have to be mindful not to complicate the exercise without enriching it by including too much data.

And finally, before getting into the details, there is one more methodological question we have been skirting around that we now need to face head on. We have used the idea of 'a city of equals' which is, for us at least, an inspiring phrase. But as we have said we do not intend to offer a model of a city of equals, or utopian blueprint, even though we have sometimes used phrases such as 'moving closer to a city of equals'. There are various reasons for our reticence. One, as should be apparent, is that we do wish to encourage the academic community and city-zens to re-examine our theory and modify it wherever necessary. For another, as in previous work (Wolff and de-Shalit 2007) we take our task to be to expose and help address inequalities, rather than provide yet another contested ideal of equality, which may or may not have connection with social policy. Discussing inequality in the absence of an articulated theory of equality, on the face of it may seem problematic, even incoherent. There are deep questions here. But those questions are for another occasion and we have discussed them elsewhere (see Wolff 2015). For practical purposes most people will agree that a society displays a highly problematic degree of inequality if, for example, schools in wealthy areas have twice the funding per pupil than schools in poorer areas. And most who agree with this claim do not have a developed theory of what funding formula or school access policy would treat every child as an equal, and may also accept that this is a question with no easy answer, for example whether funds should or should not be allocated for remedial purposes, and if so on how much and on what basis? It seems a matter of common experience to have a strong sense in particular cases that something is wrong, unequal, or unfair, and a sense for what could improve the situation, without necessarily having a view of what would be morally optimal. As Michael Walzer (1994) argues it is also interesting to note that a clear idea that something is morally wrong, unfair, or unjust, is empirically speaking often universally shared, and such a claim made in one country can be easily understood by others in other countries; whereas when we come to what is fairness, or what is just, people start raising different ideas. Accordingly, we will focus on areas where it seems manifest that there is at least potential for troubling inequalities, and where, by policy change, those inequalities can be reduced, even when there is no agreed, privileged, account of an ideal.

With these observations and caveats in place, we are now in a position to explore generating factors from the themes, and considering what sorts of measurements may be available.

6.1. Core Value 1: Non-market Access to the City's Facilities

As explained in the previous section, we feel that ultimately each city will need to decide for itself, in detailed consultation with a wide range of representative city-zens, what should be in its own index. But on the basis of our research we also feel we are in a position to make a number of recommendations about the themes they should at least take into consideration in their deliberations. That said we do not think our work is definitive and we would be delighted to hear of possibilities we have overlooked. In what follows, therefore, what we say in relation to the themes should be taken in the spirit of a non-exhaustive checklist. We would be surprised if the themes we discuss were ignored entirely, even though we accept that in some cases they may ultimately be rejected as inappropriate or too marginal for a particular city. How the factors are defined precisely, how they are measured, and how much weight they are given, will be deliberative decisions for each city.

The first theme of non-market access to the city's facilities is potentially very comprehensive and wide ranging. Broadly speaking each city would want to come up with an account of the facilities that are, or should be, available in the city, what use is being made of them, and what may explain different patterns of use, and whether any difference in use is a legitimate expression of different (group) preferences, or whether there are financial, logistical, or social barriers to their use. For example, do children from poorer families ignore tennis courts because they lack money for equipment or for court hire, or is it simply that they find football or basketball more fun? And is that because they prefer one sort of physical challenge to another, or because they want to be with their friends, who play football or basketball because they don't think of tennis as a sport for 'people like me'?

But before we get into questions of patterns of use, we need to have an account of the type of facilities that we believe should be made available to all, even if, in practice, different levels of service may be available to those with more money. Facilities that come first to mind, and have been mentioned in previous chapters, include food shops, markets, pharmacies, and health clinics, for both physical and mental health. Also important are schools, including those for children with special needs, as well as colleges of further education, leisure services such as museums, cinemas, pubs, cafes, parks, playgrounds, swimming pools, tennis courts, and football grounds; and public amenities, from public toilets to libraries, and access to the beach or other waterfront, such as a lake or river, where there is one, as there is in most major cities. Also important is the availability of subsidized or free services, concerts, libraries, museums, children's clubs, and evening classes. Facilities for day care are very important, and also, as introduced by our interviews, night care too for children whose parents or other carers work nights, though we suspect most cities would currently fail badly on this measure.

Transport, of course, is vital for access to facilities, as well as for work, and should be assessed for how frequent, varied, safe, comfortable, accessible, and affordable it is. We also would include cycling (including publicly owned bicycles) and walking, with the caveat noted above that it is more appropriate in some cities than others. Hence it is likely to require detailed investigation in each city to devise an appropriate set of measures around transport. We must not forget the urban environment. How frequent and efficient is garbage collection? Are the streets kept clean? What about noise, especially at night? Are trees planted throughout the city to give shade in the summer?

However, the factor that probably weighed heaviest on the minds of our interviewees was housing. Interestingly education was mentioned less often, but this may be because there is less of a sense that it is in crisis in big cities at the present time. But, education, as well as health care and transport, will, alongside housing, be of great significance in any index. For housing we would expect issues such as rents, availability, criteria, and length of waiting lists for subsidized housing or council housing to be at the top of the agenda for most cities. But we should also mention that when there is council housing, the aesthetics both of the accommodation and its environment need to be assessed. It is often the case that cities offer council housing located near sources of air or water pollution, and which is designed to meet a basic standard of accommodation without luxury or ornament. Amsterdam's experience with the neighbourhood of Bijlmermeer is telling. In 1975 and following the independence of Suriname, it was decided that the neighbourhood would house the mass migration from Suriname. But unfortunately, their integration did not go well, and the neighbourhood deteriorated in terms of crime, poverty, alcoholism, and so on. Following a comprehensive new plan, including tearing down many buildings, renovating and building others, and in particular moving the city's main football stadium to Bijlmermeer (the Johan Cruijff ArenA), the neighbourhood became a much more attractive location, and it now hosts more than fifty thousand inhabitants from more than a hundred ethnic groups. And as for schools, assuming most are already free, we would expect that issues such as quality of education, location, and typical length of commute will all be important. For medical services, the range available, the distance to clinics, length of waiting lists and continuity of care are all likely to be important. And as we have mentioned, convenience of visiting times cannot be ignored. Recall we are interested not just in whether these facilities are present in the city, but whether everyone is able to make use of them, whatever their level of household income, and the time pressures of their jobs and caring responsibilities.

6.2. Core Value 2: A Sense of a Meaningful Life

Many of the components of a meaningful life will depend on access to the facilities of the city, and indeed this is why access for all is so important. But, as discussed in Chapter 5, there are also other components that are much less tangible which are highly fateful for people's lives. We reported that a pattern emerging from the interviews and other studies is that many urban residents nevertheless hanker for a type of communal village life in their pocket of the city, where dog walkers chat to each other, and parents and children picnic together in the local park on a sunny afternoon, but with easy access both to the vibrancy of the city centre and the calm of places with the feel of the countryside, and to feel welcome and at home in all these disparate places. Any city would do well to incorporate measures that both create the objective conditions for community, and people's subjective sense of it.

We've also already raised the question of how the need for a sense of meaningful urban life can be translated to policy. We suggested that the role of the city is to provide the basis for people together to develop the activities and relationships that give shape to both the neighbourhood and to individual life, such as giving licences to restaurants and bars and encouraging independent small businesses, to keeping the streets safe and secure, and usable by people with different physical and psychological needs, to providing community centres, night classes, good transport into the late evening, and libraries, including cultural displays, as well as larger cultural festivals. Indeed, the spirit of a passage from Anthony Crosland's *The Future of Socialism* is almost as fresh today as it was in 1956, although it doesn't explicitly mention the importance of multiculturalism:

> We need not only higher exports and old-age pensions, but more open-air cafés, brighter and gayer streets at night, later closing-hours for public houses, more local repertory theatres, better and more hospitable hoteliers and restaurateurs, brighter and cleaner eating-houses, more riverside cafés, more pleasure-gardens on the Battersea model, more murals and pictures in public places, better designs for furniture and pottery and women's clothes, statues in the centre of new housing-estates, better-designed street-lamps and telephone kiosks, and so on *ad infinitum*.
>
> **(Crosland 2006 [1956], 402–3).**

The '*ad infinitum*' for us would include opportunities to volunteer within the community, free courses on local urban history, newcomer clubs, street signs in different languages to cater for communities of immigrants and to show respect, and joining the local community garden, a shared space where residents of the neighbourhood can together maintain a garden space where they grow vegetables, flowers, and sometimes picnic together. Also important will be opportunities to take part in formal and informal political activities, including opportunities to be involved in Participatory Budgeting, which we looked at in some detail in Chapter 5. Some of these items can be measured objectively, but it also seems that the theme of a sense of a meaningful life would benefit from surveys of individual experience. We would expect all of these factors to be incorporated into a city's index.

6.3. Core Value 3: Diversity and Social Mixing

The idea of diversity ranges over a number of differences between people: race, religion, ethnicity, language, gender, sexual orientation, disability, age, and perhaps others too. The first task is simply to record the range and numbers; some contemporary cities claim to be home to more than a hundred languages. While this, naturally, can present difficulties in communication, we largely see it as a component part of a city of equals that welcomes people of all backgrounds and types, although it should be noted that cities which tend to be overlooked by immigrants because they are not affluent or because the state makes migration to it difficult, might have few languages and still be very egalitarian. So the number of languages spoken is a relevant parameter for the city's equality only if the city is a popular destination among immigrants. Again, because we believe that the main point of measuring the city's level of equality is to compare itself to previous years, we think that if a city learns that all of a sudden, the number of languages spoken in its territory has dropped, then there might be something wrong going on. It could be, for example, that a group or several groups of people felt unwanted and left.

Also, as noted in previous chapters we believe a city of equals is a 'city of many flavours', with access to lively and diverse shops, restaurants, bars, and entertainments, the presence of a variety of types of people on the street and the feeling the city is inclusive with respect to gender, age, and race and so on. A city, though, could have wide diversity, statistically speaking, but homogeneity in each neighbourhood and hence provide little in the way of social mixing. Thus, the key issue is not so much the presence of diversity but how it is manifested.

Extreme residential segregation appears to cut against the promotion of social mixing, and, as we have noted, some theorists such as Elizabeth Anderson have argued for integration. But as indicated we hesitate to follow Anderson in all particulars here and accept there can be legitimate reasons for people (especially from minoritized groups) voluntarily to choose to live with people they identify with, rather than among strangers, even if doing so would improve economic and educational opportunities, which, in any case, is not always the case. Furthermore, living side by side with others does not in itself constitute social mixing, and there are many cases in which different groups live parallel lives in the same neighbourhood, largely ignoring the existence of the other group.

Accordingly, we are sympathetic to Iris Marion Young's account of when residential segregation is more or less acceptable, or, as we would put it, compatible with the idea of a city of equals, which we discussed in Chapter 3. To summarize, the aspects to take into account, and these could, if appropriate, be incorporated into a city's index, are:

1. Is segregation the result of deliberate discrimination in housing?
2. Is knowledge of where minoritized people live common knowledge in the city and are those areas avoided and stigmatized?
3. Do members of the majority move out of these areas?
4. Is business investment lower in these areas?
5. Are these areas generally less well provided for in terms of facilities and services?
6. Is the standard of living lower in these areas?

The more a clear yes can be given to these answers, the more problematic the situation. Conversely, we would also encourage cities to pay attention to the prevalence of privileged enclaves, and especially gated communities—fortified enclaves as they have been called—or entrance-controlled shopping areas, and also to monitor the rate and consequences of gentrification.

Finally, it is worth returning to the vexed issue of income inequality. On the one hand inequality is very often measured in terms of income inequality, but on the other we are sympathetic to the argument that if a city shows less income inequality than the country in which it is located, or at least the region, then it is likely to be homogenous and exclusive, rather than inclusive. For this reason, we have suggested that a city of equals may well show more income inequality than its region, as its open nature makes less-privileged people feel welcome, or at least more able to improve their fortunes there (Alster 2022). Accordingly, while there is reason to monitor statistics about income in the city, the main material factors that affect whether a city is a city of equals is not income inequality in the city in itself, but the types of lives available to people on lower incomes. This is something that is especially picked up in the first theme—non-market access to services and facilities—but in fact runs though all four themes of our analysis.

6.4. Core Value 4: Non-deferential Inclusion

As we have explained, the idea behind this core value is that city-zens should regard themselves as equal members with others, in which, for example, any benefits they receive from the city are considered theirs by right, not privileges arbitrarily granted and easily withdrawn, and people are not required to scrape or grovel before gate-keepers, or expend excess time or effort, to get what they are entitled to. We have suggested that inclusion in this sense means that newcomers and previously marginalized people are open-heartedly included in the job market, cultural events, social initiatives, local politics, social networks, local NGOs, and so on, irrespective of gender, race, sexual orientation, disability, or any other characteristic that previously has led to discrimination or disadvantage. Statistics can give part, though not the whole, of this picture, which, we suggest, needs to be supplemented with carefully worded surveys, or even better, interviews about people's expectations and experience for themselves and others.

In this respect the semi-metaphor of 'eye contact', discussed in Chapter 4 can be put to use, both literally—is the city designed so that people can meet each other as members of communities of equals?—and metaphorically. In this latter sense we are especially interested in patterns of asymmetrical respect: do you find yourself needing to show respect to others who show no respect to you (or, perhaps even more problematically, vice versa).

Inclusion can be a quantitative concept, such as the numbers or percentages of the newcomers or the previously marginalized and excluded who are now taking part in planning, deliberation, politics, social activities, cultural activities, or whatever parameter we can think of. (And this can be measured either subjectively, that is to say, by people's reports, or objectively, by comparing statistics and so on). But we are equally interested in inclusion as a qualitative concept, and in particular the quality of relationships. This again can be found by interviewing people and asking them about their feelings; or by asking what are the expectations of newcomers and the hosting community members from each other? What do they think should happen? And for yourself, do you feel comfortable in your own skin, able to move freely from part to part of the city? How are you treated? Have you suffered racism, sexual harassment, or homophobia, for example?

6.5. Conclusion

This has been a relatively small book on large topic. We have tried first to motivate interest in the rather neglected question of how equality should be understood at the level of the city, rather than a nation or a state, or, indeed over the globe as a whole. In answering the question for ourselves we used a method of triangulation, combining our own reflections with our reading of the literature, and, most importantly, an extensive set of interviews, conducted over ten cities in six countries. Our contribution is to suggest that in a city of equals everyone has a secure sense of place, which consists of four core-values: non-market access to services and facilities; a sense of a meaningful life; diversity and social mixing; and non-deferential inclusion. While making a number of suggestions about how to understand each of these core values and the various themes they refer to, and how much weight should be given to each theme, we also suggest that a specific index has to be devised for each city, to reflect both its own special circumstances, and the interests and values of its city-zens, who should be consulted in the construction of the index for their city. The purpose of the index is not to rank cities in an equality index, but for a city to provide an audit of itself, and to set goals and monitor progress. We do not, however, consider that we have produced the last word on any of the matters we have introduced and discussed, but hope that this topic will attract the interest of other political philosophers, scholars of urban studies, urban planners, and empirical social scientists, to sustain but perhaps modify and improve our own work.

References

Adamczyk, Alicia. 2019. 'This Map Shows the US Cities with the Greatest Income Inequality'. CNBC. 9 October. https://www.cnbc.com/2019/10/09/this-map-shows-the-us-cities-with-the-greatest-income-inequality.html.

Agence France-Presse. 2016. Burkini Ban Suspended by Nice Court, Dismissing Claim of Public Order Risk. *The Guardian*. 2 September. https://www.theguardian.com/world/2016/sep/02/burkini-ban-suspended-nice-court-france.

Aldridge, Hannah, Theo Barry Born, Adam Tinson, and Tom MacInnes. 2015. *London Poverty Profile*. Trust for London. https://trustforlondon.org.uk/research/londons-poverty-profile-2015.

Allport, Gordon. 1954. *The Nature of Prejudice*. Oxford: Addison-Wesley.

Alroey, Itamar. 2022. 'The Influence of the Urban Structure and Social Geography in Jerusalem on the Way Jews and Palestinians Perceive Each Other in the City—the Light Rail as a Case Study'. Unpublished paper submitted as a seminar paper at the Hebrew University of Jerusalem in the course Political Theory and the City.

Alster, Tal. 2022. *Conceptualizing and Measuring Inequality in Cities*. PhD dissertation, the Hebrew University of Jerusalem.

Amin, Ash and Nigel Thrift. 2017. *Seeing Like a City*. Cambridge: Polity Press.

Anderson, Elizabeth. 2010. *The Imperative of Integration*. Princeton: Princeton University Press.

Arneson, Richard, 2013. 'Egalitarianism'. Updated 24 April. http://plato.stanford.edu/entries/egalitarianism/.

Arnestein, Sherry. 1969. 'A Ladder of Citizen Participation'. *Journal of the American Institute of Planners* 35 (4): pp. 216–24.

Atkinson, Rob and Todd Swanstrom. 2012. 'Poverty and Social Exclusion'. In *The Oxford Handbook of Urban Politics*, edited by Karen Mossberger, Susan Clarke, and Peter John. New York: Oxford University Press, pp. 333–50.

Baker, Kevin. 2005. 'Rudy Giuliani and the Myth of Modern New York'. http://kevinbaker.info/americas-mayor/.

Barak, Nir. 2020. 'Civic Ecologism: Environmental Politics in Cities'. *Ethics, Policy & Environment* 23 (1): pp. 53–69.

Barak, Nir and Avner de-Shalit. 2021. 'Urbanizing Political Concepts for Analyzing Politics in the City'. In *Research Handbook on International Law and Cities*, edited by Helmut Aust, Janne Nijman, and Miha Marcenko. Cheltenham: Edward Elgar, pp. 329–41.

Barber, Benjamin. 2013. *If Mayors Ruled the World*. New Haven: Yale University Press.

Bauböck, Rainer. 2003. 'Reinventing Urban Citizenship'. *Citizenship Studies* 7 (2): pp. 139–60.

Bauböck, Rainer. 2019. *Cities vs States: Should Urban Citizenship Be Emancipated from Nationality?* EUI Global Citizenship Observatory. 16 December. https://globalcit.eu/cities-vs-states-should-urban-citizenship-be-emancipated-from-nationality/

Baum-Snow, Nathaniel and Ronni Pavan. 2013. 'Inequality and City Size'. *The Review of Economics and Statistics* 95 (5): pp. 1535–48.

Bechor, Ori. 2023, forthcoming. *Homeless People and How They Perceive their Functionings*. MA thesis, The Hebrew University of Jerusalem.

Behrens, Kristina and Frédéric Robert-Nicoud. 2014. 'Survival of the Fittest in Cities: Urbanisation and Inequality'. *The Economic Journal* 124 (518): pp. 1371–1400.

Bell, Daniel and Avner de-Shalit. 2011. *The Spirit of Cities: Why the Identity of a City Matters in the Global Age.* Princeton: Princeton University Press.

Ben-Dahan, Adi. 2017. 'Separated vs Mixed: Studying Residence Preferences among the Ultra Orthodox Jews in Jerusalem'. Seminar Paper, Hebrew University of Jerusalem.

Ben-Porat, Guy. 2008. Policing Multicultural States: Lessons from the Canadian Model. *Policing and Society* 18 (4): pp. 411–25.

Berry, Charlie. 2021. *Homelessness in England.* Shelter. https://england.shelter.org.uk/professional_resources/policy_and_research/policy_library/homelessness_in_england_2021.

Bimkom. 2012. *The Neighbourhood of Abu Tor: A Report* (in Hebrew).

Blokland, Talja. 2017. *Community as Urban Practice.* Cambridge: Polity.

Blokland, Talja. 2023. 'Whose Public Space? What City?' A paper delivered at the Who Owns the City conference, The Hebrew University of Jerusalem.

Bobo, Lawrence, Melvin Oliver, James Johnson Jr., and Abel Valenzuela Jr. (eds). 2000. *Prismatic Metropolis: Inequality in Los Angeles.* New York: Russell Sage Foundation.

Bradley, Jennifer and Bruce Katz. 2013a. 'What America's Metropolitan Revolution Can Teach Europe'. Brookings Institution. 31 October. www.brookings.edu/articles/what-americas-metropolitan-revolution-can-teach-europe.

Bradley, Jennifer and Bruce Katz. 2013b. *The Metropolitan Revolution: How Cities and Metros Are Fixing our Broken Politics and Fragile Economy.* Washington DC: Brookings Institution Press.

Brando, Nico and Katarina Pitasse-Fragoso. 2022. 'Small in the City: The Exclusion of Children from Public Spaces'. *Justice Everywhere.* 17 October. http://justice-everywhere.org/author/nicolasbrando/.

Bratt, Rachel G. and William M. Rohe. 2007. 'Challenges and Dilemmas Facing Community Development Corporations in the United States'. *Community Development Journal* 42 (1): pp. 63–78.

Briffault, Richard. 1996. 'The Local Government Boundary Problem in Metropolitan Areas'. *Stanford Law Review* 48 (5): pp. 1115–71.

Bullard, Robert. 2009. 'Addressing Urban Transportation Equity in the United States'. In *Breakthrough Communities: Sustainability and Justice in the Next American Metropolis,* edited by Paloma Pavel. Cambridge, MA: MIT Press, pp. 49–59.

Burden, Amanda. 2021. 'NYC Chief City planner Amanda Burden on Public Space and Densification'. *Archinect.com.*21 April. https://archinect.com/news/article/98435205/nyc-chief-city-planner-amanda-burden-on-public-space-and-densification.

Caldeira, Teresa P. R. 2000. *City of Walls: Crime, Segregation, and Citizenship in São Paulo.* Berkeley, CA: California University Press.

Cannavò, Peter. 2007. *Working Landscape: Founding, Preservation, and the Politics of Place.* Cambridge, MA: MIT Press.

Carpenter, Evan S. 2014. *Identifying Cultural and Non-Cultural Factors Affecting Litter Patterns in Hickory Creek, Texas.* University of North Texas, Master of Science thesis in Applied Geography. https://digital.library.unt.edu/ark:/67531/metadc699896/m2/1/high_res_d/thesis.pdf

Casa Fluminense. 2013. 'Inequality Map'. *Casa Fluminense.* https://casafluminense.org.br/inequality-map/.

Cassiers, Tim and Christian Kesteloot. 2012. 'Socio-spatial Inequalities and Social Cohesion in European Cities'. *Urban Studies* 49 (9): pp. 1909–24.

CEC NYC. n.d. *PBNYC: The People's Budget* https://www.participate.nyc.gov/processes/pb?locale=en.

Chauvin, Juan Pablo. 2021. 'Why Does COVID-19 Affect Some Cities More Than Others? Evidence from the First Year of the Pandemic in Brazil'. http://hdl.handle.net/10419/245876.
Chetty, Raj, Nathaniel Hendren, and Lawrence F. Katz. 2016. 'The Effects of Exposure to Better Neighborhoods on Children: New Evidence from the Moving to Opportunity Experiment'. *American Economic Review* 106 (4): pp. 855–902.
Cities for Adequate Housing. 2018. 'Municipalist Declaration of Local Governments for the Right to Housing and the Right to the City'. https://citiesforhousing.org.
City of Vienna n.d. 'Gender Mainstreaming in Vienna'. https://www.wien.gv.at/english/administration/gendermainstreaming/index.html.
Clarke, Susan and Gary Gaile. 1998. *The Work of Cities*. Minneapolis: University of Minnesota Press.
Cresswell, Tim. 2009. 'Place'. In *International Encyclopedia of Human Geography*, edited by Rob Kitchen and Nigel Thrift, Vol. 8. Amsterdam: Elsevier, pp. 169–77.
Crisp, Roger. 2003. 'Equality, Priority, and Compassion'. *Ethics* 113 (4): pp. 745–63.
Crosland, Anthony. 2006 [1956]. *The Future of Socialism*. London: Constable and Robinson.
Cross, Jennifer. 2021. 'What Is a Sense of Place?' Unpublished conference paper. https://www.researchgate.net/publication/282980896_What_is_Sense_of_Place.
Cycling Without Age. 2023. 'Building Better Lives: Cycling Without Age'. https://cyclingwithoutage.org.
Damyanovic, Doris et al. 2013. *Manual for Gender Mainstreaming in Urban Planning and Urban Development*, The Municipal Department of Urban Development and Planning, Vienna. https://boku.ac.at/fileadmin/data/H03000/H85000/H85400/_TEMP_/Frauen_und_Maenner_unterwegs/FUM_englisch_kleinstens.pdf.
Davies, Alex. 2016. 'Copenhagen's New Traffic Lights Recognize and Favor Cyclists'. *Wired*. 18 February. https://www.wired.com/2016/02/copenhagens-new-traffic-lights-recognize-and-favor-cyclists.
Davis, Mike. 1997. 'The Radical Politics of Shade'. *Capitalism, Nature, Socialism* 8 (3): pp. 35–9.
Decide_Madrid n.d. Descubre la plataforma de participación ciudadana del Ayuntamiento de Madrid. https://decide.madrid.es.
Devine-Wright, Patrick. 2013. 'Think Global, Act Local? The Relevance of Place Attachments and Place Identities in a Climate Changed World'. *Global Environmental Change* 23 (1): pp. 61–9.
de-Shalit, Avner. 2018. *Cities and Immigration*. Oxford: Oxford University Press.
de-Shalit, Avner. 2020. 'Political Philosophy and What the People Think'. *Australian Philosophical Review* 4 (1): pp. 4–22.
de-Shalit, Avner. 2021. 'Amsterdam: Tolerance and Inclusion'. *Critical Review of Social and Political Philosophy* 25: pp. 742–59.
de Silva, Mariana, James Angus Fraser, and Luke Parr. 2021. 'Capability Failures and Corrosive Disadvantage in a Violent Rainforest Metropolis'. *Geographical Review* 113. DOI: 10.1080/00167428.2021.1890995.
Derudder, Ben, Michael Hoyler, Peter J. Taylor, and Frank Witlox (eds). 2011. *International Handbook of Globalization and World Cities*. Cheltenham: Edward Elgar.
Discover Amsterdam. 2023. 'Milkshake Festival'. https://www.iamsterdam.com/en/whats-on/calendar/festivals/events/milkshake-festival.
Doctors Only. 2021. 'Nearly Half of the New Doctors Are Arabs or Druze'. https://publichealth.doctorsonly.co.il/2021/09/238135/ (in Hebrew; retrieved 22 May 2023).
Dorling, Danny. 2019. *Inequality and the 1%* (third edition). London: Verso.
Dorling, Danny. 2022. 'A Letter from Helsinki'. *Public Sector Focus* July/August: pp. 12–15. https://www.dannydorling.org/?p=8977.

Doyle, Arthur Conan. 1892. The Adventure of the Copper Beeches. In *The Adventures of Sherlock Holmes*. London: George Newnes Ltd. See also http://www.jimelwood.net/students/holmes/adven1/copper_beeches.pdf

Drage, Jean. 2001. *Women in Local Government in Asia and the Pacific*. Victoria University of Wellington. https://www.ucl.ac.uk/dpu-projects/drivers_urb_change/urb_society/pdf_gender/UNESCAP_Drage_Women_Local_Government_Asia_Pacific.pdf.

DSNI. n.d. 'Dudley Street Neighborhood Initiative'. www.dsni.org.

Enos, Ryan. 2014. 'Causal Effect of Intergroup Contact on Exclusionary Attitudes'. Proceedings of the National Academy of Sciences 111 (10): pp. 3699–704.

Enos, Ryan. 2016. *The Space between Us*. Cambridge: Cambridge University Press.

Enos, Ryan and Christopher Celaya. 2018. 'The Effect of Segregation on Intergroup Relations'. *Journal of Experimental Political Science* 5 (1): pp. 26–38.

Enos, Ryan and Noam Gidron. 2016. 'Intergroup Behavioral Strategies as Contextually Determined: Experimental Evidence from Israel'. *The Journal of Politics* 78 (3): pp. 851–67.

Epting, Shane. 2023. *Urban Enlightenment: Multistakeholder Engagement and the City*. London and New York: Routledge.

Euromonitor. 2013. 'Income Inequality Ranking of the World's Major Cities'. *Euromonitor International*. 31 October. https://blog.euromonitor.com/income-inequality-ranking-worlds-major-cities/.

Euromonitor. 2017. 'The World's Largest Cities Are the Most Unequal'. *Euromonitor International*, 3 March. https://www.euromonitor.com/article/the-worlds-largest-cities-are-the-most-unequal.

European Commission n.d. 'Affordable Housing'. https://urban.jrc.ec.europa.eu/thefutureofcities/affordable-housing#the-chapter.

Eurostat. n.d. 'Is Housing Affordable?' https://ec.europa.eu/eurostat/cache/digpub/housing/bloc-2b.html?lang=en.

Fainstein, Susan. 2001. 'Inequality in Global City-Regions'. *disP-The Planning Review* 37 (144): pp. 20–5.

Fainstein, Susan. 2010. *The Just City*. Ithaca, NY: Cornell University Press.

Fawcett. 2022. ' New Data Shows We Won't See Gender Equality in Local Councils until 2077'. The Fawcett Society. https://www.fawcettsociety.org.uk/news/local-council-data-2021.

Feitelson, Eran. 1991. 'Sharing the Globe: The Role of Attachment to Place'. *Global Environmental Change* 1 (5): pp. 396–406.

Fine, Michelle, Nick Freudenberg, Yasser Payne, Tiffany Perkins, Kersha Smith, and Katya Wanzer. 2003. '"Anything Can Happen with Police Around": Urban Youth Evaluate Strategies of Surveillance in Public Places'. *Journal of Social Issues* 59 (1): pp. 141–58.

Fiss, Owen, Joshua Cohen, Jefferson Decker, and Joel Rogers (eds). 2003. *A Way Out: America's Ghettos and the Legacy of Racism*. Princeton: Princeton University Press.

Flash Eurobarometer 419. 2016. 'Quality of Life in European Cities 2015'. Flash Eurobarometer 419. https://www.gbv.de/dms/zbw/848721098.pdf.

Förster, Kirsten, Wolfgang Gerlich, Hanna Posch (PlanSinn), Manuela Bauer, Jancda Schultheiß, and Astrid Waleczka (MD-OS, Section for Gender Mainstreaming). 2021. *Gender Mainstreaming Made Easy*. The City of Vienna. https://www.gleichstellungsportal.de/wp-content/uploads/sites/2/2021/07/AC16146852.pdf.

Frankfurt, Harry. 1987. 'Equality as a Moral Ideal'. *Ethics* 98 (1): pp. 21–43.

Frankfurt, Harry. 2016. *On Inequality*. Princeton: Princeton University Press.

Frediani, Alexandre Apsan. 2015. 'Space and Capabilities: Approaching Informal Settlement Upgrading through a Capability Perspective'. In *The City in Urban Poverty*, edited by Cherlotte Lemanski and Colin Marx. London: Palgrave MacMillan, pp. 64–85.

Frediani, Alexandre Apsan and Julia Hansen (eds). 2015. *The Capability Approach in Development Planning and Urban Design*. Development Planning Unit, The Bartlett, University College London. https://www.ucl.ac.uk/bartlett/development/sites/bartlett/files/wp_173-178_special_issue_on_capability_approach_final.pdf.

Friedmann, John. 1986. 'The World City Hypothesis'. *Development and Change* 17 (1): pp. 69–83.

Froyd, Justin N. 2021. 'Unprecedented Ordinance: Amsterdam Introduces a Tourist Limit'. *Tourism Review News*, 18 July. https://www.tourism-review.com/new-tourist-limit-to-help-amsterdam-fight-overtourism-news12116.

Frug, Gerald. 2001. *City Making: Building Communities without Building Walls*. Princeton: Princeton University Press.

Frug, Gerald. 2011. 'Voting and Justice'. In *Justice and the American Metropolis*, edited by Clarissa Rile Hayward and Todd Swanstrom. Minneapolis: University of Minnesota Press, pp. 201–23.

Galbraith, Kenneth. 1958. *The Affluent Society*. Boston, MA: Houghton Mifflin Company.

Galbraith, Kenneth. 1992. *The Culture of Contentment*. Boston, MA: Houghton Mifflin Company.

Gender Equality Observatory. n.d. 'Elected Mayors Who Are Female'. *Gender Equality Observatory for Latin America and the Caribbean*. https://oig.cepal.org/en/indicators/elected-mayors-who-are-female.

Giannotti, Mariana and Pedro Logiodice. 2023. 'Mobility Injustice'. Unpublished paper.

Giffinger, Rudolf and Gudrun Haindlmaier. 2010. 'Smart Cities Ranking: An Effective Instrument for the Positioning of the Cities?' *ACE: Architecture, City and Environment* 4 (12): pp. 7–26.

Glaeser, Edward. 2012. *Triumph of the City*. New York: Penguin.

Glaeser Edward, Matt Resseger, and Kristina Tobio. 2008. 'Urban Inequality'. National Bureau of Economic Research. https://scholar.harvard.edu/files/glaeser/files/urban_inequality.pdf.

Glaeser Edward, Matt Resseger, and Kristina Tobio. 2009. 'Inequality in Cities'. *Journal of Regional Science* 49 (4): pp. 617–46.

Glasgow Centre for Population Health. 2022. *Informing and Supporting Action to Improve Health and Tackle Inequality. Participatory Budgeting*. https://www.gcph.co.uk/healthy_communities/participatory_budgeting.

Glass, Ruth (ed.). 1964. *Aspects of Change*. London: MacGibbon & Kee.

Glick, Lior. 2021. 'Commuters, Located Life Interests, and the City's Demos'. *Journal of Political Philosophy* 29 (4): pp. 480–95.

Goodhart, Charles. 1975. 'Problems of Monetary Management: The U.K. Experience'. Papers in Monetary Economics 1: pp. 1–20.

Graham, Andy Lee. 2015. 'Athens Greece Special Sidewalk for Blind People'. https://www.youtube.com/watch?v=FCL5m_KJi08.

Gray, Nolan. 2022a. 'Cancel Zoning'. *The Atlantic*. 21 June. https://www.theatlantic.com/ideas/archive/2022/06/zoning-housing-affordability-nimby-parking-houston/661289/.

Gray, Nolan. 2022b. *Arbitrary Lines: How Zoning Broke the American City and How to Fix It*. Washington, DC: Island Press.

Greek Boston. n.d. 'Why Are There Yellow Grooves in the Sidewalks of Greece?' https://www.greekboston.com/travel/grooves-sidewalks/.

Greenberg Raanan, Malka and Noam Shoval. 2014. 'Mental Maps Compared to Actual Spatial Behavior Using GPS Data: A New Method for Investigating Segregation in Cities'. *Cities* 36: pp. 28–40.

Habitat for Humanity. 2020. 'The Early History of Portland's Racist Housing Strategies: Part One'. *Habitat for Humanity Portland Region* 26 June. https://habitatportlandregion.org/the-early-history-of-portlands-racist-housing-strategies-part-one/.

Halsema, Femmke. n.d. 'Imagined Urban Communities'. European Institute. https://www.youtube.com/watch?v=1ohrh2mywUk.

Harkins, C. 2019. *An Evaluation of Glasgow City Participatory Budgeting Pilot Wards 2018/19.* Glasgow: Glasgow Centre for Population Health.

Harlena, Sharon, Anthony J. Brazela, Lela Prashada, William Stefanov, and Larissa Larsen. 2006. 'Neighborhood Microclimates and Vulnerability to Heat Stress'. *Social Science and Medicine* 63 (11): pp. 2847–63.

Harvey, David. 1985. *The Urbanization of Capital.* Baltimore: Johns Hopkins University Press.

Harvey, David. 2009 [1973]. *Social Justice and the City* (revised edition). Athens, GA: University of Georgia Press.

Harvey, David. 2019. *Rebel Cities.* London: Verso Books.

Harvey, David and Cuz Potter. 2009. 'The Right to the Just City'. In *Searching for the Just City* edited by Peter Marcuse, Peter James Connolly, Johannes Novy, Ingrid Olivo, Cuz Potter, and Justin Steil. London: Routledge, pp. 40–52.

Hayward, Clarissa and Todd Swanstrom (eds). 2011a. *Justice and the American Metropolis.* Minneapolis: University of Minnesota Press.

Hayward, Clarissa and Todd Swanstrom. 2011b. 'Introduction'. In *Justice and the American Metropolis*, edited by Clarissa Hayward and Todd Swanstrom. Minneapolis: University of Minnesota Press, pp. 1–29.

Hernandez, Bernardo, M. Carmen Hidalgo, M. Esther Salazar-Laplace, and Stephany Hess. 2007. 'Place Attachment and Place Identity in Natives and Non-Natives'. *Journal of Environmental Psychology* 27 (4): 310–19.

Hood, Suzanne. 2004. 'Reporting on Children in Cities: The State of London's Children Reports'. *Children, Youth and Environment* 14 (2): pp. 113–23.

Imrie, Rob, Loretta Lees, and Mike Raco (eds). 2009. *Regenerating London: Governance, Sustainability and Community in a Global City.* London: Routledge.

Issar, Ya'ara. n.d. *Abu Tor.* The Jerusalem Institute for Policy Research. Research report no. 474. (In Hebrew).

Jacobs, Jane. 1961. *The Death and Life of Great American Cities.* New York: Random House.

Johansson, Håkan and Alexandru Panican (eds). 2016. *Combating Poverty in Local Welfare Systems: Active Inclusion Strategies in European Cities.* Basingstoke: Palgrave Macmillan.

Karssenberg, Hans, Jeroen Laven, Meredith Glaser, and Mattijs van 't Hoff (eds). 2016. Second version. *The City at Eye Level.* Un-Habitat. Delft: Eburon. https://thecityateyelevel.com/app/uploads/2018/06/eBook_The.City_.at_.Eye_.Level_English.pdf.

Kaddar, Merav. 2020. 'Gentrifiers and Attitudes towards Agency: A New Typology: Evidence from Tel Aviv-Jaffa, Israel'. *Urban Studies* 57 (6): pp. 1243–59.

Khanna, Parag. 2012 'The Cities-States: Interview with Florian Rittmeyer'. 5 April. https://www.paragkhanna.com/the-cities-state/.

King, Loren. 2011. 'Public Reason and the Just City'. In *Justice and the American Metropolis*, edited by Clarissa Hayward and Todd Swanstrom. Minneapolis: University of Minnesota Press, pp. 59–81.

Kirchberg, Volker. 2015. 'Museum Sociology'. In *Routledge International Handbook of the Sociology of Art and Culture*, edited by Lauri Hanquinet and Mike Savage. London: Routledge, pp. 232–46.

Klinenberg, Eric. 2002. *Heat Wave: A Social Autopsy of Disaster in Chicago.* Chicago: University of Chicago Press.

Knobel, Lance. 2014. 'Berkeley Ranks 10th for Income Inequality among U.S. Cities'. *Berkeleyside.* 28 April. https://www.berkeleyside.com/2014/04/28/berkeley-ranks-10th-for-income-inequality-among-u-s-cities.

Kohn, Margaret. 2011. 'Public Space in the Progressive Era'. In *Justice and the American Metropolis*, edited by Clarissa Hayward and Todd Swanstrom. Minneapolis: University of Minnesota Press, pp. 81–104.

Kohn, Margaret. 2013. 'What Is Wrong with Gentrification?' *Urban Research and Practice* 6 (3): pp. 297–310.

Kohn, Margaret. 2016. *The Death and Life of the Urban Commonwealth*. New York: Oxford University Press.

Knight Foundation and Gallup. n.d. *Soul of the Community*. Knight Foundation. https://knightfoundation.org/sotc/.

Lees, Loretta, Tom Slater, and Elvin Wyly. 2008. *Gentrification*. New York: Routledge.

Lenard, Patti Tamara. 2013. Making politics more welcoming. *The Literary Review of Canada*. https://reviewcanada.ca/magazine/2013/06/making-politics-more-welcoming/

Lenard, Patti Tamara. 2015. Residence and the right to vote. *Journal of International Migration and Integration*. 16 (1): pp. 119–32

Levine, Daphna and Meirav Aharon. 2022. The Social Deal: Urban Regeneration as an Opportunity for In-place Social Mobility. *Planning Theory* (online first) https://doi.org/10.1177/14730952221115872.

Light, Andrew. 2003. 'Urban Ecological Citizenship'. *Journal of Social Philosophy* 34 (1): pp. 44–63.

Litman, Todd. 2023. 'Evaluating Transportation Equity: Guidance for Incorporating Distributional Impacts in Transport Planning'. *Victoria Transport Policy Institute*. 16 April. https://www.vtpi.org/equity.pdf.

Long, James E., David W. Rasmussen, and Charles T. Haworth. 1977. 'Income Inequality and City Size'. *The Review of Economics and Statistics* 59 (2): pp. 244–6.

Loukaitou-Sideris Anastasia and Renia Ehrenfeucht. 2012. *Sidewalks: Conflict and Negotiation over Public Space*. Cambridge, MA: MIT Press.

Löw, Martina. 2012. 'The Intrinsic Logic of Cities: Towards a New Theory on Urbanism'. *Urban Research & Practice* 5 (3): pp. 303–15.

Löw, Martina. 2013. 'The City as Experiential Space: The Production of Shared Meaning'. *International Journal of Urban and Regional Research* 37 (3): 894–908.

Lu, Wei and Alexander Tanzi. 2019. 'In America's Most Unequal City, Top Households Rake in $663,000'. *Bloomberg Economics*. 21 November. https://www.bloomberg.com/news/articles/2019-11-21/in-america-s-most-unequal-city-top-households-rake-in-663-000.

Macedo, Stephen. 2011. 'Property Owning Plutocracy and American Localism'. In *Justice and the American Metropolis*, edited by Clarissa Hayward and Todd Swanstrom. Minneapolis: University of Minnesota Press, pp. 33–58.

Mahadevia, Darshini and Sandip Sarkar. 2012. *Oxford Handbook of Urban Inequalities*. Oxford: Oxford University Press.

Magnusson, Warren, 2011. *Politics of Urbanism. Seeing Like a City*. London: Routledge

Marcuse, Peter. 1985. 'Gentrification, Abandonment, and Displacement: Connections, Causes, and Policy Responses in New York City'. *Journal of Urban and Contemporary Law* 28: pp. 195–240.

Marcuse, Peter. 1997. 'The Enclave, the Citadel, and the Ghetto: What Has Changed in the Post-Fordist U.S. City'. *Urban Affairs Review* 33 (2): 228–64.

Marcuse, Peter, Peter James Connolly, Johannes Novy, Ingrid Olivo, Cuz Potter, and Justin Steil (eds). 2009. *Searching for the Just City*. London: Routledge.

Markovits, Daniel. 2019. *The Meritocracy Trap*. London: Penguin.

Marsh, Ben, Allan Parnell, and Ann Joyner. 2010. 'Institutionalization of Racial Inequality in Local Political Geography'. *Urban Geography* 31 (5): pp. 691–709.

Maslow, A. H. 1943. 'A Theory of Human Motivation'. *Psychological Review* 50 (4): pp. 370–96.

Massey, Douglas S. and Nancy A. Denton. 1988. 'The Dimensions of Residential Segregation'. *Social Forces* 67 (2): pp. 281–315.

McManus, Kat. 2020. 'Planning for Inclusivity: How Vienna Built a Gender-equal City'. Local Government Information Unit. https://lgiu.org/planning-for-inclusivity-how-vienna-built-a-gender-equal-city/#:~:text=Vienna%20has%20carried%20out%20more,and%20 alleyways%20by%20adding%20mirrors (retrieved 12 May 2023).

McLaren, Duncan and Julian Agyeman. 2015. *Sharing Cities: A Case for Truly Smart and Sustainable Cities*. Cambridge, MA: MIT Press.

Meagher, Sharon. 2007. 'Philosophy in the Streets: Walking in the City with Engels and de Certeau'. *City* 11 (1): pp. 7–20.

Miller, Lee J. and Lu Wei. 2019. 'These Are the Most Expensive U.S. Cities, Based on Sushi Prices'. *Bloomberg UK*, 23 July. https://www.bloomberg.com/news/articles/2018-07-23/sushinomics-show-tuna-rolls-sting-seattle-cheap-in-new-orleans?leadSource=uverify %20wall.

Moore, Margaret. 2019. 'The Taking of Territory and the Wrongs of Colonialism'. *Journal of Political Philosophy* 27 (1): pp. 87–106.

Musterd, Sako. 2006 Segregation, Urban Space, and the Resurgent City'. *Urban Studies* 43 (8): pp. 1325–40.

Musterd, Sako, Szymon Marcińczak, Maarten van Ham, and Tiit Tammaru. 2017. 'Socioeconomic Segregation in European Capital Cities: Increasing Separation between Poor and Rich'. *Urban Geography* 38 (7): pp. 1062–83.

Musterd, Sako and Wim Ostendorf. 2012. 'Inequalities in European Cities'. In *International Encyclopedia of Housing and Home*, edited by Susan J. Smith, Marja Elsinga, Lorna Fox O'Mahony, Ong Seow Eng, Susan Wachter, and David Clapham, Vol. 4. Amsterdam: Elsevier, pp. 49–55.

National Bureau of Economic Research. n.d. *Moving to Opportunity*. http://www2.nber.org/mtopublic/.

National Democratic Institute n.d. *The Woman Mayors' Network (WoMN)*. https://www.ndi.org/women-mayors-network-womn.

National Geographic. n.d. *Finding Urban Nature*. https://education.nationalgeographic.org/resource/finding-urban-nature.

Nijman, Jan and Yehuda Dennis Wei. 2020. 'Urban Inequalities in the 21st Century Economy'. *Applied Geography* 117: pp. 102–88.

NPR. 2022. 'A Day Trip to Venice Will Require a Reservation—and a Fee'. NPR. 3 July. https://www.npr.org/2022/07/03/1109615164/italy-venice-travel-new-rules.

Nussbaum, Martha. 2000. *Women and Human Development*. Cambridge: Cambridge University Press.

Nussbaum, Martha. 2011. *Creating Capabilities: The Human Development Approach*. Cambridge, MA: Harvard University Press.

O'Connor, Alice, Chris Tilly, and Lawrence Bobo. 2001. *Urban Inequality: Evidence from Four Cities*. New York: Russell Sage Foundation.

OECD. 2018. *Divided Cities: Understanding Intra-Urban Inequalities*. OECD. https://www.oecd.org/publications/divided-cities-9789264300385-en.htm.

Okamoto, Dina and Kim Ebert. 2016. 'Group Boundaries, Immigrant Inclusion, and the Politics of Immigrant–Native Relations'. *American Behavioral Scientist* 60 (2): pp. 224–50.

Ostow, Robin. 2005. 'Mokum Is Home: Amsterdam's Jewish Historical Museum'. *European Judaism: A Journal for the New Europe* 38 (2): pp. 43–68.

Parfit, Derek. 1997. 'Equality and Priority'. *Ratio* 10 (3): pp. 202–21.

Parnell, Susan (2015) 'Poverty and "the City"'. In *The City in Urban Poverty*, edited by Charlotte Lemanski and Colin Marx. London: Palgrave McMillan, pp. 16–39.

Pavel, Paloma (ed.). 2009. *Breakthrough Communities: Sustainability and Justice in the Next American Metropolis.* Cambridge, MA: MIT Press.

People's Palace Projects and Redes da Maré. 2020. *Building the Barricades* https://peoplespalaceprojects.org.uk/wp-content/uploads/2021/11/folder-infografico-ing-online-5.pdf.

Pettigrew, Thomas and Linda Tropp. 2006. 'A Meta-Analytic Test of Intergroup Contact Theory'. *Journal of Personality and Social Psychology* 90 (5): pp. 751–78.

Pettit, Philip. 2014. *Just Freedom: A Moral Compass for a Complex World.* New York: W. W. Norton.

Phillips, Anne. 1985. *Divided Loyalties.* London: Virago.

Polner, Robert (ed.). 2005. *America's Mayor: The Hidden History of Rudy Giuliani's New York.* Brooklyn, NY: Soft Skull Press.

Powell John. 2009. 'Reinterpreting Metropolitan Space as a Strategy for Social Justice'. In *Breakthrough Communities: Sustainability and Justice in the Next American Metropolis* edited by Paloma Pavel. Cambridge, MA: MIT Press, pp. 23–33.

Pritchard, John, Diego Bogado Tomasiello, Mariana Giannotti, and Karst Geurs. 2019. 'Potential Impacts of Bike-and-Ride on Job Accessibility and Spatial Equity in São Paulo, Brazil'. *Transportation Research Part A: Policy and Practice*, Vol. 121: pp. 368–400.

Rae, Douglas. 2011. 'Two Cheers for Very Unequal Incomes: Toward Social Justice in Central Cities'. In *Justice and the American Metropolis*, edited by Clarissa Hayward and Todd Swanstrom. Minneapolis: University of Minnesota Press, pp. 105–24.

Reardon, Sean and Kendra Bishoff. 2011. 'Income Inequality and Income Segregation'. *American Journal of Sociology* 116 (4): 1092–153.

Rosenzweig, Roy and Elisabeth Blackmar. 1992. *The Park and the People: A History of Central Park.* New York: Cornell University Press.

Saaby, Tina. 2015. 'Copenhagen: A City for People'. Lecture. https://www.youtube.com/watch?v=tr5Aluh-P30.

Sadik-Khan, Jannette. n.d. 'New York Streets? Not So Mean Anymore'. Ted Lecture. http://www.jsadikkhan.com/media.html.

Sampson, Robert J. and William Julius Wilson. 1995. 'Toward a Theory of Race, Crime, and Urban Inequality'. In *Crime and Inequality*, edited by John Hagan and Ruth D. Peterson. Stanford, CA: Stanford University Press, pp. 37–56.

Sampson, Robert J., William Julius Wilson, and Hanna Katz. 2018. 'Reassessing "Toward a Theory of Race, Crime and Urban Inequality": Enduring and New Challenges in 21st Century America'. *Du Bois Review* 15: pp. 13–34.

Sandel, Michael J. 2012. *What Money Can't Buy: The Moral Limits of Markets.* New York: Farrar, Straus and Giroux.

Sassen, Saskia. 1991. *The Global City:* New York, London, Tokyo. Princeton: Princeton University Press.

Sassen, Saskia. 1999. *Globalization and its Discontents: Essays on the New Mobility of People and Money.* New York: New Press.

Schaeffer, Katharine. 2022. 'Key Facts about Housing Affordability in the U.S'. Pew Research Center, 23 March. https://www.pewresearch.org/fact-tank/2022/03/23/key-facts-about-housing-affordability-in-the-u-s/.

Schlichtmann, John Joe, Jason Patch, Marc Lamount Hill. 2018. *Gentrifier.* Toronto: University of Toronto Press.

Schragger, Richard. 2013. 'Is a Progressive City Possible? Reviving Urban Liberalism for the Twenty-First Century', *Harvard Law & Policy Review* 7: pp. 231–52.

Schultz, Westley, Renee Bator, Lori Brown Large, Coral M. Bruni, and Jennifer Tabanico. 2013. 'Littering in Context: Personal and Environmental Predictors of Littering Behavior'. *Environment and Behavior* 45 (1): pp. 35–9.

Sen, Amartya. 1992. *Inequality Re-examined.* Cambridge, MA: Harvard University Press.
Sen, Amartya. 1999. *Development as Freedom.* Oxford: Oxford University Press.
Sen, Amartya. 2009. *The Idea of Justice*, London: Allen Lane.
Sevano, Victoria. 2022. 'Amsterdam Proposes New Rules to Protect Housing from Investors'. *I Am Expat.* 17 February. https://www.iamexpat.nl/housing/real-estate-news/amsterdam-proposes-new-rules-protect-housing-investors#:~:text=Amsterdam%20has%20already%20announced%20that,money%2C%E2%80%9D%20the%20municipality%20says.
Shanahan, Danielle, Richard Fuller, Robert Bush, Brenda Lin, and Kevin Gaston. 2015. 'The Health Benefits of Urban Nature: How Much Do We Need?' *BioScience* 65 (5): pp. 476–85.
Shany, Ayelet. 2022. 'No Doubt; We Are Betraying Our Children'. An interview with Ram Cohen. *Haaretz.* 22 June.
Shelby, Tommie. 2014. 'Inequality, Integration, and Imperatives of Justice: A Review Essay'. *Philosophy & Public Affairs* 42 (3): pp. 253–85.
Shelby, Tommie. 2016. *Dark Ghettos: Injustice, Dissent, and Reform.* Cambridge, MA: Harvard University Press.
Shmaryahu-Yeshurun, Yael. 2022. 'Retheorizing State-led Gentrification and Minority Displacement in Global South-east. *Cities.* 130. https://doi.org/10.1016/j.cities.2022.103881.
Silver, Christopher. 1997. 'The Racial Origin of Zoning in American Cities'. In *Urban Planning and the African American Community: In the Shadows*, edited by June Manning Thomas and Marsha Ritzdorf, pp. 23–39. Thousand Oaks, CA: Sage Publications, pp. 23–39.
Simmel, Georg. 1903. 'The Metropolis and Mental Life'. In *The Sociology of Georg Simmel*, translated by Kurt Wolff. New York: Free Press, 1950, pp. 409–24.
Smith, Richard. 2013. 'Beyond the Global City Concept and the Myth of "Command and Control"'. *International Journal of Urban and Regional Research* 38 (1): pp. 98–115.
Soderstrom, Mary. 2008. *The Walkable City.* Montreal: Véhicule Press.
Soja, Edward. W. 2010. *Seeking Spatial Justice.* Minneapolis: University of Minnesota Press.
Stone, Clarence N. 2006. 'Power, Reform, and Urban Regime Analysis'. *City and Community* 5 (1): pp. 23–38.
Sugrue, Thomas. 1996. *The Origins of Urban Crisis.* Princeton: Princeton University Press.
Sundstrom, Ronald. 2013. 'Comment on Elizabeth Anderson's *The Imperative of Integration*'. *Symposia on Gender, Race and Philosophy* 9 (2). https://sgrp.typepad.com/sgrp/fall-2013-symposium-anderson-on-integration.html.
Suss, Joel and Thiago Oliveira. 2022. 'Economic Inequality and the Spatial Distribution of Stop and Search: Evidence from London'. *The British Journal of Criminology.* https://doi.org/10.1093/bjc/azac069.
Swedish Institute. 2022. 'Taxes in Sweden'. Swedish Institute. 6 May. https://sweden.se/life/society/taxes-in-sweden.
Tan, Khee Giap, Wing Thye Woo, Kong Yam Tan, Linda Low, and Grace Ee Ling Aw. 2012. *Ranking the Liveability of the World's Major Cities: The Global Liveable Cities Index.* Singapore: World Scientific Publishing
Tawney, R. H. 1931. *Equality.* London: George Allen and Unwin.
Terrill, William and Michael Reisig. 2003. 'Neighbourhood Context and Police Use of Force'. *Journal of Research in Crime and Delinquency* 40 (3): pp. 291–321.
Therborn, Göran. 2009. 'Why and How Place Matters'. In *The Oxford Handbook of Contextual Political Analysis*, edited by Robert E. Goodin and Charles Tilly. Oxford: Oxford University Press, pp. 509–33.
Tilly, Charles. 1998. *Durable Inequality.* Berkeley and Los Angeles: University of California Press.
Tonkiss, Fran. 2015. *Divided Cities.* Lecture at the LSE. Divided Cities: Urban inequalities in the 21st Century—YouTube https://www.youtube.com/watch?v=r4iXe5l4whY.

Tonkiss, Fran. 2017. 'Urban Economies and Social Inequalities'. In *The SAGE Handbook of the 21st Century City*, edited by Suzzane Hall and Ricky Burdett. London: Sage, pp. 187–200.
Townsend, Peter. 1979. *Poverty in the United Kingdom*. London: Penguin.
Trust for London. n.d. *London Poverty Profile*. https://www.trustforlondon.org.uk/data.
Uitermark, Justus. 2009. 'An *In Memoriam* for the Just City of Amsterdam'. *City* 13 (2–3): pp. 347–61.
UN-Habita, 2009. *Planning Sustainable Cities*. https://unhabitat.org/planning-sustainable-cities-global-report-on-human-settlements-2009
UN-Habitat, 2016. *The Widening Urban Divide*. https://www.un-ilibrary.org/content/books/9789210582810.
Urban Nature Atlas. 2023. *Welcome to Urban Nature*. https://una.city/.
Valentine, Gill. 2008. 'Living with Difference: Reflections on Geographies of Encounter'. *Progress in Human Geography* 32 (3): pp. 323–37.
Van Bochove, Marianne. 2012. *Geographies of Belonging: The Transnational and Local Involvement of Economically Successful Migrants*. Rotterdam: Erasmus University Press.
Van Eijk, Gwen. 2010. *Unequal Networks, Spatial Segregation, Relationships, and Inequality in the City*. Amsterdam: IOS Press.
Van Leeuwen, Bart. 2010. 'Dealing with Urban Diversity: Promises and Challenges of City Life for Intercultural Citizenship'. *Political Theory* 38 (5): pp. 631–57.
Van Leeuwen, Bart. 2018. 'To the Edge of the Urban Landscape: Homelessness and the Politics of Care'. *Political Theory* 46 (94): pp. 586–610.
Van Leeuwen, Bart. 2020. 'What Is the Point of Urban Justice? Access to Human Space'. *Acta Politica* 57: pp. 169–90.
Verloo, Mieke. 1999. *Gender Mainstreaming: Practice and Prospects*. Report prepared for the Council of Europe. EG (99) 13. 7 January. https://rm.coe.int/1680596141.
Verloo, Mieke, 2000. 'Making Women Count in the Netherlands'. In *Making Women Count: Integrating Gender into Law and Policymaking*, edited by Sue Nott, Fiona Beveridge, and Kylie Stephen. Farnham: Ashgate, pp. 49–77.
Vieira, Tuca, 2017. 'Inequality . . . in a Photograph'. *The Guardian*. 29 November. https://www.theguardian.com/cities/2017/nov/29/sao-paulo-injustice-tuca-vieira-inequality-photograph-paraisopolis.
Waldron, Jeremy. 1991. 'Homelessness and the Issue of Freedom'. *UCLA Law Review* 39 (1): pp. 295–324.
Walker, Julian. 2013. 'Time Poverty, Gender and Well-being: Lessons from the Kyrgyz Swiss Swedish Health Programme'. *Development in Practice* 23: pp. 57–68.
Wander Women Project. n.d. 'The Future of Gender Representation on Traffic & Road Signs'. https://wanderwomenproject.com/the-future-of-gender-representation-on-traffic-road-signs/.
Walzer, Michael. 1983. *Spheres of Justice*. New York: Basic Books.
Walzer, Michael. 1994. *Thick and Thin: Moral Argument at Home and Abroad*. Notre Dame, IN: University of Notre Dame Press.
Weinstock, Daniel. 2011. Self-Determination for (Some) Cities? In *Arguing about Justice: Essays for Philippe Van Parijs*, edited by Axel Gosseries and Philippe Vanderborght. London: UCL Press, pp. 377–86.
Weinstock, Daniel. 2014. 'Cities and Federalism'. In *Nomos 55, Federalism and Subsidiarity*, edited by James Fleming and Jacob Levy. Cambridge: American Society for Political and Legal Philosophy, pp. 259–90.
Whyte, William. 1980. *The Social Life of Small Urban Spaces*. Washington, DC: The Conservation Foundation.
Williamson, Thad. 2013. 'Mobility and Its Opponents: Richmond, Virginia's Refusal to Embrace Mass Transit'. Unpublished.

Wilson, William Julius. 2012 [1987]. *The Truly Disadvantaged: The Inner City, the Underclass, and Public Policy* (second edition). Chicago: University of Chicago Press.

Wittgenstein, Ludwig. 1953. *Philosophical Investigations*. Oxford: Blackwell Publishing.

Wolff, Jonathan. 1998. 'Fairness, Respect, and the Egalitarian Ethos'. *Philosophy & Public Affairs* 29 (2): pp. 97–122.

Wolff, Jonathan. 2006. 'Risk, Fear, Shame, Blame and the Regulation of Public Safety'. *Economics and Philosophy* 22 (3): pp. 409–27.

Wolff, Jonathan. 2015. 'Social Equality and Social Inequality'. In *Social Equality: Essays on What It Means to Be Equals*, edited by Carina Fourie, Fabian Schuppert, and Ivo Wallimann-Helmer. Oxford: Oxford University Press, pp. 209–26.

Wolff, Jonathan. 2017. 'Forms of Differential Social Exclusion'. *Social Philosophy and Policy* 34 (1): pp. 164–85.

Wolff, Jonathan. 2019a 'Equality and Hierarchy'. Proceedings of the Aristotelian Society 119 (1): pp. 1–23.

Wolff, Jonathan. 2019b [2011]. *Ethics and Public Policy*. London: Routledge.

Wolff, Jonathan. 2020. 'Public Reflective Disequilibrium'. *Australasian Philosophical Review* 4 (1): pp. 45–50.

Wolff, Jonathan and Avner de-Shalit. 2007. *Disadvantage*. Oxford: Oxford University Press.

Wolff, Jonathan and Avner de-Shalit. 2013. 'On Fertile Functionings: A Response to Martha Nussbaum'. *Journal of Human Development and Capabilities* 14 (1): pp. 161–5.

Wolman, Harold, assisted by Robert McManmon. 2012. 'What Cities Do: How Much Does Urban Policy Matter?' In *The Oxford Handbook of Urban Politics* edited by Karen Mossberger, Susan Clarke, and Peter John. Oxford: Oxford University Press, pp. 415–42.

Wright, Erik Olin. 2010. *Envisioning Real Utopias*. London: Verso.

Yorukoglu, Mehmet. 2002. 'The Decline of Cities and Inequality'. *American Economic Review* 92 (2): 191–7.

Young, Iris Marion. 1990. *Justice and the Politics of Difference*. Princeton: Princeton University Press.

Young, Iris Marion. 2000. *Inclusion and Democracy*. Oxford: Oxford University Press.

Young, Iris Marion. 2011. *Responsibility for Justice*. Oxford: Oxford University Press.

Young, Michael and Peter Willmott. 2011 [1957]. *Family and Kinship in East London*. London: Routledge.

Yujelevski, Tanya. 2023. 'Real Estate Prices in Berlin 2022: Statistics & Evolution'. Sweet Home. 25 January. https://www.sweet-home.co.il/en/real-estate-prices-in-berlin-2022-statistics-evolution/#:~:text=real%20estate%20prices%20in%20germany&text=The%20 real%20estate%20market%20in,of%20about%2010%25%20per%20year.

Žlender, Vita and Stefano Gemin, 2020. Testing Urban Dwellers' Sense of Place: Towards Leisure and Recreational Peri-urban Green Open Spaces in Two European cities. *Cities*. 98. https://doi.org/10.1016/j.cities.2019.102579.

Zukin, Sharon. 1995. *The Culture of Cities*. Malden, MA: Wiley-Blackwell.

Zweig, Stepfan. 1943. *The World of Yesterday*. New York: Viking Press.

Index

For the benefit of digital users, indexed terms that span two pages (e.g., 52–53) may, on occasion, appear on only one of those pages.